Medicine and Biomedical Sciences in Modern History

Series Editors

Carsten Timmermann
University of Manchester
Manchester, United Kingdom

Michael Worboys
University of Manchester
Manchester, United Kingdom

The aim of this series is to illuminate the development and impact of medicine and the biomedical sciences in the modern era. The series was founded by the late Professor John Pickstone, and its ambitions reflect his commitment to the integrated study of medicine, science and technology in their contexts. He repeatedly commented that it was a pity that the foundation discipline of the field, for which he popularized the acronym 'HSTM' (History of Science, Technology and Medicine) had been the history of science rather than the history of medicine. His point was that historians of science had too often focused just on scientific ideas and institutions, while historians of medicine always had to consider the understanding, management and meanings of diseases in their socio-economic, cultural, technological and political contexts. In the event, most of the books in the series dealt with medicine and the biomedical sciences, and the changed series title reflects this. However, as the new editors we share Professor Pickstone's enthusiasm for the integrated study of medicine, science and technology, encouraging studies on biomedical science, translational medicine, clinical practice, disease histories, medical technologies, medical specialisms and health policies. The books in this series will present medicine and biomedical science as crucial features of modern culture, analysing their economic, social and political aspects, while not neglecting their expert content and context. Our authors investigate the uses and consequences of technical knowledge, and how it shaped, and was shaped by, particular economic, social and political structures. In re-launching the Series, we hope to build on its strengths but extend its geographical range beyond Western Europe and North America. **Medicine and Biomedical Sciences in Modern History** is intended to supply analysis and stimulate debate. All books are based on searching historical study of topics which are important, not least because they cut across conventional academic boundaries. They should appeal not just to historians, nor just to medical practitioners, scientists and engineers, but to all who are interested in the place of medicine and biomedical sciences in modern history.

More information about this series at
http://www.palgrave.com/gp/series/15183

María Jesús Santesmases

The Circulation of Penicillin in Spain

Health, Wealth and Authority

palgrave
macmillan

María Jesús Santesmases
Instituto de Filosofía
Consejo Superior de Investigaciones
Científicas
Madrid, Spain

Medicine and Biomedical Sciences in Modern History
ISBN 978-3-319-69717-8 ISBN 978-3-319-69718-5 (eBook)
https://doi.org/10.1007/978-3-319-69718-5

Library of Congress Control Number: 2017959319

Printed on acid-free paper

This Palgrave Macmillan imprint is published by Springer Nature
The registered company is Springer International Publishing AG
The registered company address is: Gewerbestrasse 11, 6330 Cham, Switzerland

Acknowledgements

A considerable debt of gratitude is accumulated when conducting a lengthy research project. My interest in the history of penicillin began many years ago, when I discovered a file at the Rockefeller Archive Centre on the Spanish Professor of Physiology at the University of Madrid Medical School, Antonio Gallego. This material, and advice from Margarita Salas, led me to his son, Roberto Gallego; our discussions provided me with insightful observations of the early days of penicillin in Spain through the history of *Compañía Española de Penicilinas y Antibióticos* (CEPA). For my reconstruction of this company's history I received generous help from Santiago López and Antonio Gómez de Mendoza. I learnt about Banco Urquijo and its enterprises at seminars on the history of industry led by Gómez de Mendoza at the Fundación Zubiri in Madrid, though I certainly did not have the opportunity to learn all the group could have taught me about industry during Franco's regime. Francisco Comín provided early publishing support, accepting a working paper to be distributed by Fundación Empresa Pública. Later, Viviane Quirke organised an inspiring meeting in Oxford on twentieth-century pharmaceuticals, at which I presented a first attempt to reconstruct the history of the firm in an international perspective. Following her useful advice I was able to contribute a chapter dealing with the origins of this work to a book edited by Viviane and Judy Slinn. Being a member of the European research programme DRUGS between 2008 and 2013, awarded by the European Science Foundation (06-RNP-045) and funded by national funding agencies, enabled me to keep in close contact with colleagues tackling similar research issues through that period. As part of this programme, Miguel García-Sancho, Ana Romero de Pablos, Christoph Gradmann and I organised an international workshop on the history of antibiotics in Madrid; a special issue of *Dynamis* including many of the contributions was then produced, edited by Christoph and myself. Teresa Ortiz, Agata Ignaciuk, and I organised a DRUGS meeting on Gendered Drugs in Granada, at which I presented my first study on women in the history of antibiotics, and together with Teresa edited a volume on the subject published by Ashgate, through which I was able to study and develop a gender perspective. At a meeting in Berlin organised by Mathias Grote and Lara Keuck I presented a paper on the history of the bacterial cell wall, a reconstruction of the experimental research carried out on the effects of antibiotics on bacteria, part of which is also included here.

At various archives I received kind and invaluable assistance from many librarians and archivists, including at the Archivo General de la Administración, in Alcalá de Henares (Madrid); Archivo del Ministerio de Hacienda (Madrid); the Rockefeller Archive Center in North Tarrytown (New York); The British Library Additional Manuscripts section (London); the Fondos Banco Urquijo, at the library of Instituto Juan March de Investigaciones Sociales (Madrid), later moved to the Instituto Carlos III-Juan March, Universidad Carlos III de Madrid; and the Hemeroteca Nacional at the Biblioteca Nacional de España (Madrid). Without the continued support of librarians at the Biblioteca Tomás Navarro Tomás, CSIC (Madrid), this book would not have been possible.

Previous versions of this book's contents, that I have since rearranged, reworked and developed, have been published in *Antibióticos en la Autarquía* (Madrid: Fundación Empresa Pública, 1999); 'Distributing penicillin: the clinic, the hero and industrial production in Spain, 1943–1952', in *Perspectives on Twentieth-Century Pharmaceuticals*, ed. Viviane Quirke and Judy Slinn (Oxford-Berna: Peter Lang, 2010), 91–118; 'Screening antibiotics: industrial research in CEPA and Merck in the 1950s', *Dynamis* 31 (2011): 407–27; 'Gender in research and industry: women in antibiotic factories in 1950s Spain', in *Gendered Drugs and Medicine: Historical and Socio-cultural Perspectives*, ed. T. Ortiz-Gómez and M. J. Santesmases (Farnham: Ashgate, 2014), 61–84; 'The long postwar and the politics of penicillin: early circulation and smuggling in

Spain, 1944–1954', *Medicina nei Secoli* 26 (2014): 615–38; and 'The bacterial cell wall in the antibiotic era: an ontology in transit between morphology and metabolism, 1940s–1960s', *Journal of the History of Biology* 49 (2016): 3–36. I was also able to participate in a collective reflection about the history of antibiotic resistance, for which my thanks go to Scott Podolsky and Anne Kveim Lie.

Throughout the process of elaborating my ideas on including additional agents to the somewhat circumscribed history of antibiotic discovery and early antimicrobial practices, I am indebted to many colleagues and friends whose insights, discussions and comments about their own research and mine helped me develop my story and refocus my interests. Esteban Rodríguez Ocaña carefully and patiently read most of the chapters, providing insightful comments and criticisms, and along with Ramón Castejón, shared unpublished material with me on penicillin and venereal diseases in Spain. I have been able to explore the gender approach with my inspiring colleague and friend, Teresa Ortiz, and benefitted from discussions with Ana Romero de Pablos and Agata Ignaciuk. Mauro Capocci told me that Fleming's diary of his trip to Spain in 1948 was included in The Alexander Fleming Papers at the British Library. Many others provided additional material, testimonies and suggestions: I am particularly grateful to Jon Arrizabalaga, Ana Barahona, Robert Bud, Ramón Castejón, Mar Cebrián, Francisco Comín, Angela Creager, Flurin Condrau, Rosa García Ballester, Pedro García Barreno, Antonio Gómez Mendoza, Christoph Gradmann, Santiago López, Ilana Löwy, Jorge Martínez Reverte, Emilio Muñoz, Maribel Porras, Toine Pieters, Vivianne Quirke, Maria Rentetzi, Edna Suárez, and Ulrike Thoms.

I am also indebted to the generous information provided in the many interviews that Roberto Gallego, Isabel Martín, Sagrario Mochales, as well as Fernando Peláez and Carlos Rosas, granted me on CEPA's history. As witnesses and major characters in this story, their recollections have been highly instrumental in my reconstruction. Encouraging support from Michael Worboys and Carsten Timmermann enabled me to transform my initial project draft into this book. An anonymous referee made insightful comments and suggestions that helped me enhance both my arguments and the originality of my narrative. I am especially grateful to the expert and committed help provided by Joanna Baines in copy-editing my English; her skills prompted me to elaborate some of my arguments further.

The research for this book received funding from the Comunidad de Madrid (Madrid regional Government) and the Spanish ministry of Economy and Innovation (FFI2012-34076 and FFI2016-76364).

Antonio Calvo Roy provided the daily support my long research endeavour required. Our children, Atocha, Manuel and Carmen, have grown up alongside this venture and provided comments and suggestions, helping keep the project in perspective.

Although this text has benefitted enormously from all those who participated in the construction of my approach—the historiography included—and in its expression, I am solely responsible for any mistakes in, or omissions from, the text that follows.

Madrid
July 2017

Contents

List of Figures

CHAPTER 1

Introduction: The West, Spain and the Early Circulation of Penicillin

In this chapter I explore the political and material context of a post-war Spain suffering the tragic loss of civil rights and material goods under the Franco dictatorship. I review the early days of penicillin, utilising material published by fellow historians of science and medicine, such as the pioneering contributions of Gladys Hobby and Robert Bud. The circumstances surrounding the first batches of penicillin to reach Spain will be narrated, along with initial public revelations about its miraculous properties. The main objective of this book will then be outlined: to reflexively explore the historiographical arguments—discussed throughout the manuscript—on the circulation of penicillin through time and space during the long post-war.

In March 1944, the Madrid daily newspaper *ABC* reported that penicillin had been used in Spain for the first time 'with magnificent results'. A new medicine 'that is lacking here' had arrived. The father of a severely ill little girl posed with his sister for a press agency photo distributed the day after being taken, holding in his hands a parcel of the drug to treat his young daughter. The next day an earlier use was reported: a young mining engineer in La Coruña, a sea port in North-West Spain, had been injected with penicillin for acute septicaemia: 'The first injection was applied yesterday at 11 am, when he had a temperature of 39 degrees … today before a second dose he was at 37', a clear improvement.[1] Penicillin had reached Spain through Gibraltar, the British harbour on the South coast of Spain, shipped from an 'African port occupied by the Allies'.[2]

M. J. Santesmases, *The Circulation of Penicillin in Spain*, Medicine and Biomedical Sciences in Modern History, https://doi.org/10.1007/978-3-319-69718-5_1

This penicillin that had entered Spain, the public fame of its therapeutic capacity, the methods to obtain and purify it and the protocols for its clinical use, composed a set of material, social and rhetorical practices embedded in the early days of the Franco dictatorship and the immediate post-Second World War years. The policies of Franco and the Allies interacted to disseminate this new drug during the long post-war period: the long aftermath of the Spanish Civil War (1936–1939) and the Second World War. Protracted negotiations authorised transportation of the drug from the United States to Madrid, and provided protocols for use, while black market penicillin circulated through unofficial channels.

Along with most of Europe, Spain had joined in the heroic reception afforded this new wonder drug since at least 1943, when penicillin first appears in Spanish archival material and newspapers. This heroic representation was based on a miraculous capacity to cure previously fatal infections: effects greeted as marvels by an impoverished population in the devastation of the wars and during the first decade of Franco's dictatorship. Accounts of such healing crossed national and cultural boundaries effortlessly; as had infection, ingrained in memories across Europe, regardless of education, social position and age, since the 1918 Spanish flu pandemic.[3] While infection rates increased during the Spanish Civil War, as they did during the Second World War, the treatment of war wounds improved. New techniques were in use, including a method for the treatment of broken bones devised by the Spanish orthopaedic surgeon, Josep Trueta.[4] Blood transfusion techniques were also developed and practised during the Civil War: the Barcelonian haematologist, Frederic Duran i Jorda, was one of the front's pioneers, while the Canadian physician Norman Bethune gained renown by organising the first ever mobile blood transfusion service, to support the Republican Government.[5] The internationalisation of the War of Spain—as it was known in the foreign, left-wing environments which provided the International Brigades—was part of a Zeitgeist that bisected political ideologies following the Communist revolution. During the convulsive, critical and promising interwar period, Europe suffered effects from the deep economic crisis in the USA in 1929 at a time when the continent was still reeling from the Versailles treaty marking the end of the Great War.

While the Spanish Civil War produced technical and medical improvements embedded in a tragically restrictive environment for invention and experimentation, the public health system was dismantled. A culture of urgency revolving around solving problems related to the state of emergency was initiated, and indeed constituted, by the war itself. The circularity that characterises war regimes, however, included the permanent embodiment of an everyday tragedy: the loss of people and spaces, geographical and emotional, shared in time, and anxieties such as prolonged poverty and a lack of basic food and refuge. The war was self-nourishing, fed by every triumph and tragedy of lives lost.

Among the survivors, infections were—and would remain until the present time—one of the faces of war-generated poverty. Reducing the prevalence of infections was a measure of well-being, directly correlated with public health since the early days of hygiene and vaccination in the late nineteenth century.[6] Bacteriological doctrine and the possibility of producing specific vaccines and sera for every germ responsible for disease played a significant part in the way such infections were conceptualised, managed and encountered.[7] Smallpox, tuberculosis and poliomyelitis were among those most successfully prevented by vaccination after long research and preventive campaigns. Hygiene campaigns mobilised social and medical resources to provide preventive measures, with an extended conception of infections and pandemics as avoidable if proper measures were taken. Yet infections were often still difficult to avoid, stop or cure. By 1935 a new drug on the market was believed to have the ability to—as John Lesch has phrased it—keep some infectious disease 'under control'.[8] This was Prontosil, a sulfa-drug produced by the chemical research industry: a red dye effective against a number of infections in mice, and which proved in human beings to kill the germ without damaging the host. A turning point in the industrialisation of medicine, the chemistry of dyestuffs was at the origins of the new chemotherapeutic agent, a transformative medicine marking the shift from natural products to industrially manufactured synthetics. The capabilities of chemistry during the first third of the twentieth century also included the chemical weapons successfully used during the Great War. This set of successful and tragic events constituted the basis upon which chemistry assumed its mantle as the twentieth-century agent of a new scientific and technological regime.

Reports of the amazing new antimicrobial penicillin travelled through time and space.[9] Like sulfa-drugs, the novelty of penicillin related to many domains of activity: research, industry and medicine. However, in each of those domains penicillin revealed its singularity. Here was a drug isolated from microorganisms, therefore a return to natural products, but rather than originating from plants, penicillin came from the very life forms it was intended to destroy. Living beings had become the landscape in which to discover new therapeutic agents.

Rumours of wondrous effects when applied to wounds were some of the earliest celebrated successes of penicillin. Reports spread swiftly through post-Civil War Spain. As the Second World War began, Franco's army had already entered Madrid, where his entry of the capital would represent victory and mark the end of the war. News of this wonder drug spread through the war regime of circulation: along with reports on battles, losses and casualties travelled information of cures performed by the yellow powder of unpurified penicillin.

These overlapping influential temporalities must be taken into account when reconstructing the history of the medical and scientific object that penicillin became in all its forms and derivatives. While the Second World War was manufacturing new medical and technical objects, including atomic weapons, the Spanish population had no opportunity to recover from the devastation of the Civil War. Poverty, hunger, infections and the terror imposed by Franco contributed to the misery of life that historians Antonio Cazorla, Conxita Mir and Ángela Cenarro have revealed, in the process of rejuvenating historiographical approaches to post-war Spain through local histories with wide-ranging impact.[10] The Spanish government retained food ration cards through the post-Civil War period, right up until the early 1950s, and a thriving black market of basic foodstuffs developed under the repressive policies of the early dictatorship. Death rates due to undernourishment and infections, which had declined during the early 1930s, increased not only during the Civil War but throughout the first half of the 1940s: it was not until 1946 that infections lost their position as the primary cause of death in Spain.[11] During this decade, known as the Years of Hunger, the precariousness of urban life was compounded by insalubrious water.[12] Meanwhile the dictatorship covered up epidemics of polio and typhus, considering them issues of law and order: equating medical infection with ideological contagion.[13]

Along with many other republican and left-wing physicians, scientists and intellectuals, Josep Trueta went into exile, living in Britain until his retirement. Trueta was among the first witnesses in Oxford of the anti-infectious action of early samples of impure penicillin, and would contribute to circulating information about the wonders of the drug when writing to colleagues and relatives who remained in Barcelona after the war.[14] Knowledge of such wonders travelled faster than the drug itself.

The effects of penicillin would be experienced by troops at the front during the Second World War. Although Spain did not participate, it felt the conflict's impact in many ways. With such a colossal amount of death and illness, poverty and destruction, the Second World War became a shared enterprise contributing to the creation of a new geopolitical order. Suffering and learning combined over the six years it lasted, consolidating transnational ways of sharing knowledge and policies, in which penicillin became a symbolic and medical commodity with significance well beyond its healing capacity. The materiality of the drug was longed for everywhere that knowledge of its capacity to cure had reached. Cure as a desire, as an expected salvation, condensed in penicillin's healing capacity. It was an object received in Spain from abroad, somewhere across the ocean, far from the miseries of everyday post-war life, as a lifesaver. The mobility of information about it participated in the powers of the drug; 'networks of travels, inquiry and change' were involved in the boundary crossing of penicillin's fame.[15]

Spanish newspapers first acknowledged penicillin in October 1943, relaying a report about John Mahoney, Director of the Marine Hospital on Staten Island, New York, and his successful treatment with penicillin of four syphilitic patients.[16] In Spain, venereal disease was diagnosed and treated in 'dispensaries', with sufferers prescribed penicillin from the 1950s on. Considering how fast knowledge about penicillin travelled and the fact batches of the drug had officially arrived in Madrid by 1945, this was a considerable delay. Venereal diseases were some of the earliest infections to be successfully treated in the USA with the new drug during the Second World War, therefore clinical experts in dermatology—associated due to the manifestation of syphilis in skin sores—were some of the first to be aware of the drug's miraculous effects.[17] While news of syphilis being treated with Salvarsan (see below) had appeared in Spanish papers during the 1910s, the successful use of penicillin against such a feared infection went unreported in coverage of the drug's miraculous effects during the 1940s.[18] This silence reflected attitudes to public hygiene under the dictatorship, as a moral and ideological approach to infections led to sexually transmitted disease being ignored. Patients were subject to a twofold moral opprobrium: being infected, and regarded as having behaved reprehensibly to become infected. The authorities had no interest in prevention.

It took time for penicillin from the pharmaceutical companies in the USA and UK that manufactured the drug to reach the wounded and infected. Penicillin's biography, despite its healing capacities being celebrated throughout a world of war, poverty and destruction was, by 1946, still as astoundingly short as the drug was strikingly effective.

From a Petri Dish to the Front

Histories of twentieth-century medical drugs have underlined the role of penicillin as a miraculous commodity since its creation in the 1940s, as an antimicrobial obtained from a mould; also a microbe.[19] Penicillin became an emblematic wonder drug of the post-Second World War years and the history of its discovery, production and distribution have been celebrated for half a century.

Anti-infectious chemotherapy preceded penicillin, however, with the arsenic-based drug Salvarsan proving active against syphilis, one of the most lethal known infections. This had been identified during the first decade of the twentieth century in the German dyestuff industry, with Paul Ehrlich emerging as the chemical hero of the very first *Zauberkugel*, or magic bullet. As promising as it appeared, Ehrlich's Salvarsan produced toxic reactions. Gerhard Domagk's Prontosil unlocked the field of sulfa-drugs, the wonders and limitations of which would provide a culture of chemotherapy that helped change perceptions of infectious disease, its treatment and the medical knowledge these new practices of healing would inform. Streptococcal infections, pneumonia, meningitis, dysentery, urinary tract infections and the sexually transmitted infection, gonorrhoea, were cured by Prontosil. This anti-infection sulfonamide was able to destroy particular parasites while leaving the host's cells intact.[20]

Both Prontosil and Salvarsan produced a culture of infection cure, of expectancy, in a way preparing society for the arrival of penicillin. Arsenic aniline derivatives synthesised at Ehrlich's laboratory were manufactured by Hoecht, while Bayer produced the sulfonamide provided by Domagk: both firms belonged to Germany's powerful IG Farben conglomerate of dyestuff industries. Expertise in organic chemistry, and in dye manufacturing—the origins of industrial organic compound production—created the first substances active against syphilis—Salvarsan, and later the less toxic derivative Neo-salvarasan—while Prontosil proved active against staphylococcal infections. Despite the expectations raised by these healers of familiar and dreaded diseases, their success was limited and toxicity high; it was lack of toxicity and broad spectrum activity that fuelled the huge success of penicillin and would be among the principal properties sought for in the testing of any new antimicrobial throughout the antibiotic era.

Also associated with the particular chemical research agenda from which sulphonamide and arsenic-based anti-infective drugs emerged, penicillin participated in changing public cultures of science: scientific research would come to be regarded as the basis of medical activity and source of new therapeutic products. Many diseases would now be cured if research was promoted, in both academic and industrial research laboratories. Scientific knowledge would be achieved as a research enterprise; the more intensively pursued and financially supported the better for cure of disease and the health of society.

During the Second World War anti-infectious therapy shifted from chemistry to microbiology. Microbes would be the source of the new drugs, the chemical activity of which would keep microbiological research occupied for decades. Microbiology as a knowledge space would join chemistry in the task of producing and identifying new antimicrobials. This biopolitical landscape had its own historicity, from the chemical era to one involving microbes as sources of new antimicrobial drugs: microbes killing microbes would be the phenomena to be sought, witnessed and selected for the isolation of new anti-infection products for decades to come. This shift also indicates a geographical move of pharmaceutical manufacturing from German chemical leadership in anti-infective drug detection and industrial manufacturing up until the Second World War, when pharmaceutical firms in the United States, and slightly later in the UK, took over; although German expertise in chemistry would remain one of the bases for the country's post-war industrial recovery. It is the USA, however, which stands out as the nation whose use and production of penicillin have been comprehensively studied.[21]

The paper in which Fleming described penicillin—having detected antimicrobial effect in a Petri dish of a mould named *Penicillium notatum*—was published in 1929 and followed by exploratory trials.[22] As a bacteriologist at a renowned London hospital for infectious diseases, and having trained with the British vaccine expert, Almroth Wright, at St Mary's Hospital in London, Fleming was transformed into a hero when, a decade later, Howard Florey, Ernst Chain and Norman Heatley at Oxford University's Department of Pathology extracted a therapeutic agent from the mould in tiny amounts, enabling them to test its effects in ten cases.[23]

Florey, Chain and Heatley, together with Edward Abraham, Arthur D. Gardner, the two women among the signatories, Margaret Jennings and Jean Orr-Ewing, both MDs from Oxford University, and Arthur G. Sanders published their report in 1940, and information about their culturing methods—how to make *Penicillium* grow and the kind of culture vessels to use—the following year.[24] Microbiologists in small laboratories across Europe attempted to culture the mould in their Petri dishes.[25] Methods and tests were provided in scientific, academic publications, and in internationally distributed specialised journals, making the information they included available to every subscribing university, research laboratory and research institution, as well as those that exchanged journals; almost every research institution created their own journal, and swapping with others was routine.

The fascination generated by these first few successes in curing infectious wounds led to the mould extract—a yellow powder—being brought to the USA in a journey funded by the Rockefeller Foundation, after Florey had failed to convince British pharmaceutical firms of the substance's capabilities. By then private funding for US research in the natural sciences was substantial and powerful, in a system that had based its wealth on a liberal approach to arts, science and industry. It was during the Second World War at the Northern Regional Research Laboratory (NRRL) of the US Department of Agriculture in Peoria (Illinois) that an efficient method of production was devised. As Gladys Hobby has phrased it, 'America evaluated penicillin's potential'. The few successfully treated cases were a tiny sample on which to base the 'potential' this new drug embodied, which manifested in the early cases of bacterial endocarditis Hobby herself treated with penicillin cultures in patients at New York's Presbyterian Hospital in 1942.[26]

Testing in the large fermentation tanks at the new NRRL included the addition of corn-steep liquor—a by-product of maize starch wet milling—to the fermentation medium and submerging of the mould culture growth. Promotion of 'new uses of farm products' was part of a strategy by the US department of Agriculture at Peoria.[27] At the top political level, support from Alfred Newton Richards, vice-president of the Committee of Medical Research of the Office of Scientific Research and Development (OSRD), promoted penicillin production as part of the US war effort. Production moved from the pilot plant to industrial manufacturing through the involvement of pharmaceutical firms such as Merck, Pfizer, Squibb and Lilly. The effectiveness of infection treatment with penicillin was confirmed through the treatment of wounds, surgical infections and venereal diseases at the front.[28]

This method of penicillin production was patented by the NRRL's Andrew Moyer; knowledge regarding production details thus produced a secondary commodity.[29] As a by-product of penicillin and its effective cure of some infections, protection of information on how to manufacture it, and the rarity of corn-steep liquor in countries lacking the immense maize industry of the USA, made penicillin production in far-flung settings—the European ones—initially difficult. Secrets of research for the war effort and the early participation of industry made the extraction of penicillin a protected process: protected for the US manufacturing firm from the expectations it generated among European manufacturers, medical practice and the general population, informed of its effects and fascinated by its success. Although every known infection would not be cured by the new drug, the expectation produced by its success became a promise beyond penicillin's actual capabilities, overstated by a devastated continent.

The Second World War and penicillin became interconnected during the early days of this medical cure, its use and the distribution policy of the magic substance that cured familiar, deadly infections. It was a medicine produced by the US war effort of, and for, the needs of war, a weapon that would be useful—according to Alfred Newton Richards, director of the US council coordinating the war effort in medical research—for 'military medicine'. The year 1943 is still regarded as pivotal in the circulation of penicillin's public renown, a time when the drug was being produced by a handful of US factories run by pharmaceutical firms. According to Richards's report that year, penicillin should be used principally for the US Army overseas and the war wounds of soldiers returning from the Pacific front. Although industrially manufactured, the drug was still scarce.[30]

In August 1943, the US National Research Council's Committee on Chemotherapy, chaired by Chester Keefer, from the Boston University School of Medicine, published its first report on the treatment of 500 patients with penicillin manufactured by Merck, Squibb and Pfizer. They investigated the infections most likely to be suffered by the Army and those resistant to sulphonamides, and concluded that the antimicrobial drug was 'notably potent'. Through the British Medical Research

Council, North African campaigns in Algiers became the scenario for one of the first successful treatments of infections with penicillin on the Allied war front in the summer of 1943.[31] Together with Brigadier Hugh Cairns, Howard Florey went to Algiers and personally supervised tests on war wounds, with units of penicillin he carried from London, manufactured in Britain. Florey's long report on the effects of penicillin in the African Allied front was published in October: 'Because the supply was scanty and likely to be so for some time, the emphasis was on local, rather than systemic, administration, e.g., as a powder or in a solution.'[32] Results in recent wounds were encouraging, but treatment of older injuries was not as effective. During the subsequent Allied invasion of Sicily, with more experience of the drug and its application, treatment 'led to outstanding results'.[33]

In 1944, when mass-production of penicillin in the USA had begun, not only for the treatment of war wounds, but also for wider clinical trials, the Comité Nacional del reparto de la penicilina en España (National Committee for the Distribution of Penicillin in Spain) was established. The scarcity of the drug, together with its effectiveness, made penicillin both a desirable and expensive product. Its use preceded manufacture almost everywhere except the USA. Clinicians in Spain reported its wondrous effects from 1944 onwards, and a number of medical researchers attempted to produce it through a surface culture.

Clinicians and the Public

Large-scale US production of penicillin began in May 1943 after promising clinical evaluations by the American Committee of Medical Research and with support from the War Production Board.[34] In Spain, the Comité Nacional was charged with both distributing the batches of penicillin received from the USA and establishing criteria for approving or rejecting requests.[35] Carlos Jiménez Díaz, a highly respected physician and clinical researcher, was appointed chairman of the Comité Nacional, which counted amongst its members clinician Eduardo Ortiz de Landázuri. Jiménez Diaz had been diagnosed with pneumonia during the summer of 1944 and, at the suggestion of his colleagues and family, was successfully treated with penicillin only a few weeks before the arrival of the first legal batch from the USA.[36]

According to a report published in 1945, as a result of early contacts made by the Comité Nacional, a protocol describing which diseases should be treated with penicillin and which should not was received at the end of 1944; the protocol had been sent by Keefer.[37] At that time US penicillin production was increasing and civilian investigations of the antibiotic in treating certain infectious diseases had begun. From July 1943 on, Keefer was in charge of all American supplies.[38] The Comité Nacional may well have been inspired by the Penicillin Panel of the American National Research Council, while a shortage of the drug on legal markets also contributed to enhancing the fame of penicillin in Spain and its perception as practically magical.

During the second half of the 1940s, penicillin was obtainable in Spain from the Comité Nacional or, more often, through the black market.[39] Fictional and non-fictional accounts of the post-Second World War period in Spain contain stories of penicillin being bought at—among other places—a well-known Madrid public house, and the fishing port of a seaside town.[40] The medical perception of this new drug was constructed on the basis of both its therapeutic properties and its scarcity.

The academic network of clinicians and medical researchers at that time embraced the new therapeutic object with enthusiasm. Among the earliest detailed information about penicillin was a book published in 1945 by the respected physician, José Álvarez-Sierra, *Lo que cura la Penicilina. Presente y porvenir de una droga mágica*. A well-known medical writer and practitioner, Álvarez-Sierra gained his knowledge of the drug from a 'huge amount of journals, publications, monographs and proceedings of academic sessions' that he had received 'despite the difficulties of the mail'.[41] It had been through Latin American colleagues that Álvarez Sierra first received news of penicillin: an Argentinian clinician who had spent time in Pennsylvania; a colleague from Panama at the Mayo Clinic in Rochester; another from Sao Paulo. Requests for information on the drug from his Spanish colleagues, according to the preface of *Lo que cura la Penicilina*, prompted Álvarez-Sierra to write the book. Following preliminary descriptions of fungi and bacteria, the book deals with penicillin activity, dosages and the conditions it could cure, alongside results from early clinical reports. Álvarez-Sierra's book is a useful account of the knowledge available at that time, presented in a comprehensible manner. The most active propagandist for the new curative product was Florencio Bustinza. This British-born Spaniard, a pharmacist and natural scientist by training, began his career as a high school teacher and in 1943 was appointed professor at the University of Madrid's Natural Sciences Faculty. In 1944 he published his first article on the discovery of penicillin for a lay audience in the widely distributed Madrid daily *ABC*.[42] He also published articles and two books, and gave many presentations on antibiotics during the second half of the 1940s, beginning in 1945 with a talk on the history of penicillin and antibiotics.[43]

Bustinza's visit to the USA from August 1945 to December 1946 took him to some of the main laboratories and facilities involved in the early production of penicillin. This was the kind of trip supported by the USA and the Spanish Junta de Relaciones Culturales—a board under the Ministry of Foreign Affairs, which granted awards for foreign travel to scientists and professionals—that helped create the public fame of US industry, research activities and technological developments, constructed following Allied victory in the Second World War with US support. Many Spanish clinicians and scientists travelled there while being in touch with José Romañá, a Spanish Jesuit and mathematician with close ties to authorities at the research institution created by Franco in 1939, the Consejo Superior de Investigaciones científicas (CSIC).[44] Romañá kept a watchful eye on the Spaniards who travelled to the USA with the support of the Junta de Relaciones Culturales, carefully reading the reports they provided and requesting more information when he considered it necessary.

During his visit to the USA, Bustinza visited NRRL, where the effective submerged culture of the mould had been developed. He also met with Henry Welch at the FDA laboratory for the control of penicillin in Washington, and with Keefer at Boston's Evans Memorial Hospital. Bustinza met a number of other scientists involved in penicillin research, including Leo Loewe at the Jewish Hospital in Brooklyn, who used penicillin to treat bacterial endocarditis, and Harry Eagle, who administered it for syphilis at Johns Hopkins University, Baltimore.

Bustinza also visited several US factories already manufacturing penicillin, observing first-hand the wide network of pharmaceutical firms providing the drug to the world. He visited the factories of Commercial Solvents Corporation in Terre Haute, Indiana; Merck in Rahway, New Jersey; Abbot in North Chicago; Lederle in Pearl River, New York; Squibb in New Brunswick, New Jersey; and Wyeth at West Chester, Pennsylvania. His travels reveal widely-distributed manufacturing by an immense country with powerful technologies, able to both produce the miracle drug and efficiently exhibit the processes to those in need.

Bustinza's trip ended in New York when he embarked for Europe on December 11, 1945. He reached Southampton, in England, on December 17, and three days later visited Alexander Fleming at Saint Mary's Hospital in London. They would meet many times from then on. Bustinza's nine-year-old daughter had been cured by penicillin some months earlier, when conjunctivitis in her right lachrimal gland had caused general inflammation: according to his memoirs 'a few intramuscular injections were enough'.[45] Access to penicillin was a privilege that helped maintain an established social order: the Comité Nacional was the only official delivery agent. The drug could fetch a high price on the black market, making it available to the wealthy, and unreachable for the poor. Scarcity consolidated the social order supporting Franco's early dictatorship and steadfastly contributed to its stability: in the hands of the Spanish authorities that governed its distribution and were well informed about the black market, penicillin became an additional tool for social and political control.

Bustinza's account in his 1945 book, *De Pasteur a Fleming*, had boosted public knowledge of penicillin's effects in the treatment of infections, as well as the perception of Fleming's work as a heroic achievement. He kept in personal contact with Fleming after his visit to Britain in 1945, and following Fleming's death a decade later, published a book about their relationship. According to one of Fleming's biographers, Bustinza became 'one of Fleming's most ardent, and certainly eloquent, admirers' and played a major part in creating the heroic perception in Spain of both the new drug and its discoverer. His tone was attuned to that which Franco claimed for himself and his supporters in the victorious rebellion against the Second Spanish Republic.[46]

By the time Bustinza published his first article in *ABC*, many physicians were already aware of the therapeutic success of penicillin. Between late 1943 and mid-1944, as early information on the beneficial effects of the new drug were beginning to circulate, a group of clinicians and chemical researchers in Barcelona repeatedly attempted to develop a method for obtaining penicillin by culturing the mould *Penicillium notatum*. By growing a mould sample obtained from the Lister Institute in London, they discovered ways of extracting the antibiotic and testing it without access to literature on the methods previously used. Despite their limited skills with the new drug, they reported success testing penicillin in dogs and mice, demonstrating the modest demands of microbiological practices, which took place in simple laboratory facilities, with the microscope and a trained microbiological eye as main participants.[47]

Bustinza, being an English-speaking biologist, performed as a literal translator. Learning English was not hugely popular among educated Spaniards, who tended towards French and German. But Bustinza also became a symbolic translator of the power of the drug, the fascinating biological wonders of a mould and its secrets, the action in Petri dishes and in the patient, through his public involvement with the media.

In 1945, reviews were published in Spanish medical journals in which the therapeutic activity of penicillin against infections was reported, as well as the drug's limitations.[48] By that time, members of the Comité Nacional were already contributing to professional knowledge on the type of infections for which penicillin treatment was recommended. Protocols for its medical use were becoming known, and the image of penicillin as a magical, wonderfully useful and effective drug was being widely publicised.[49]

Successful treatments at the hands of Spanish physicians were soon reported. A case of septicaemia caused by staphylococci was cured early in 1945, the haemoculture becoming negative within two days. To achieve this, 450,000 units of 'American penicillin' had been used after 'going through the necessary formalities'.[50] Such formalities might include either a request to the Comité Nacional, or contact with black market traders. Between late 1944 and June 1945, more than a thousand patients were treated with penicillin by clinicians 'from all around Spain' in collaboration with the Comité Nacional.[51] For many families the cure at that time of a young infected child, friend or neighbour is a cherished memory even today.[52]

In 1946, the otorhinolaryngologist Mateo Jiménez Quesada reported sixty cases of successful treatment of throat, nasal and lower respiratory tract infections using penicillin. He treated sinusitis caused by Pfeiffer's bacillus and sent his results to Alexander Fleming, who, according to Jiménez Quesada's recollections, had stated that this bacillus was not sensitive to penicillin. By that time, Fleming was obtaining similarly good results against the same type of infection by administering the drug by inhalation. Thus, Fleming suggested to Jiménez Quesada that 'by inhalation one can get in the sputum a concentration sufficient to kill [the bacillus]'.[53] Jiménez Quesada had also travelled to the USA in 1945 to learn about progress made with antibiotic treatments. He visited hospitals and research laboratories, among them the Presbyterian Hospital in New York and the Mayo Clinic in Rochester, where he witnessed penicillin being successfully used.[54] Jiménez Quesada himself would be the main character in a news reel by Noticiarios y Documentales, the state-controlled producer of news and documentary cinema reels to promote the policies of Franco's dictatorship, generally known as No-Do. A young boy and an impoverished old woman are shown being treated with penicillin through inhalation in what appears to be a charity hospital. No name was pronounced in the documentary, only images showing the active agency of penicillin in curing sinusitis, along with a display of the doctor's medical skills, a couple of woman nurses and the inhalator in a consulting room and by a hospital bed. The boy and old woman were the patient representatives selected for the documentary: innocent children and the poverty-stricken elderly. An earlier report by No-Do on penicillin in June 1945 was dedicated to its use by Allied forces; the end of the Second World War was nearing, and the drug and the Allied side had both attracted the Spanish government's attention, despite Franco having explicitly supported the Axis powers.[55]

Penicillin occupied all spaces; social, political and medical. It was found in the daily press, in No-Do documentaries, in academic discourse, and in fiction. The promises it created were an inherent part of the way capitalism produces wealth: by creating futures for business, by capturing virgin markets and by captivating cultures of salvation from the debris of war. Penicillin's ability to cross political and geographical boundaries is comparable to that of its crossing of disciplinary ones. From one country to another, across shared cultures of infection and emerging policies of hygiene, the wonders of penicillin circulated easily, swiftly and dramatically. Associated with tragic death and dramatic cure, it became a commodity that helped stabilise the 'irresistible empire' of consumption that Victoria de Grazia has analysed.

But the history of such commodities as the group of drugs penicillin became proved to be endless. Embedded in the fascination of twentieth-century achievements in research and biomedicine, the materials used to confront infections included both diagnosis and treatment, and antibiotics became embodied in both; as did such other biomedical technologies of the twentieth century, such as X-rays. But despite these research products creating such promising technological regimes—penicillin was considered the main character in the history of infection cure for decades—and becoming agents of healing and faith in science and technology, it would soon become apparent that nature and biology, in the form of bacteria and germs in general, were able to mount a resistance to these expectations.

So while the circulation of antibiotics contributed to the internationalisation of the wonders and expectations created by penicillin, this was largely due to the internationalisation of infections. Infections circulated widely and effortlessly, global epidemics crossed political and geographical frontiers and, during the colonial era, became tools of empires, infecting natives, preserving conquest, and transforming religious gatherings into powerful culturing media.[56] Knowledge about penicillin and its wonders crossed the most impenetrable national and political boundaries, turning frontiers into crossing borders, bridging gaps as, previously, infections had. But this ability to circulate knowledge about penicillin was based on a shared culture of knowing: shared—along with commodities—between East and West, colony and metropolis. The reception of penicillin was not only a matter of knowledge however, but also practices. The drug's materiality was involved in this process at a time when the flow of commodities was not as straightforward as that of knowledge; the circulation of the commodity itself involved a wide diversity of agents. Trade, politics and the law intervened in the production, and governed the movements, of penicillin ampoules.

Thus, paper technologies and their distribution became an instrumental agent in the expansion of a culture: books, letters, images in drawings and photographs became sources and means of geographically widening anything recently witnessed, described, tasted or tested. But it was trade—economic and political authority—that intervened in the circulation of commodities of any kind, whether medicines or materials. Penicillin as knowledge and practice is therefore a chance to explore, to test the role played by journeys, movements and transmissions in the construction of contemporary cultures and lifestyles. General Franco barely travelled abroad, but scientists, industrialists and traders did, from and to Spain. As I will explore in the following chapters, these exchanges of knowledge and practices intervened by creating a culture in which penicillin became an agent in contemporary history.

Notes

1. 'La medicina y los medicos', *ABC*, March 12, 1944, 34.
2. *ABC Sevilla*, March 12, 1944, 10.

3. On the pandemic in Spain, see M. I. Porras Gallo, 'La lucha contra las enfermedades "evitables" en España y la pandemia de la gripe de 1918–19', *Dynamis* 14 (1994): 159–83; M. I. P. Gallo, *Un reto para la sociedad madrileña: la epidemia de gripe de 1918–19* (Madrid: Editorial Complutense, 1997).
4. Josep Trueta, *Fragments d'una vida. Memories* (Barcelona: Edicions, 1978), 62. On Trueta, see J. L. Barona, 'Un exili de transició. Científics republicans a Gran Bretanya', *Mètode* (2012): 72–9.
5. On health and the Spanish Civil War see J. L. Barona and E. Perdiguero-Gil, 'Health and the war. Changing schemes and health conditions during the Spanish civil war', *Dynamis* 28 (2008): 103–26. On Duran, Josep-Eladi Baños, 'Un pionero', *El País*, February 11, 2006.
6. Among the historiography of vaccination, see Michael Worboys, *Spreading Germs: Disease Theories and Medical Practice in Britain, 1865–1900* (Cambridge: Cambridge University Press, 2000).
7. Isabel Porras Gallo, Maria José Báguena and Mariano Ayarzagüena, eds., *La erradicación y el control de las enfermedades infecciosas* (Madrid: Catarata, 2016); Stuart Blume, Christine Holmberg and Paul Greenough, eds., *The Politics of Vaccination. A Global History* (Manchester: Manchester University Press, 2017).
8. John E. Lesch, *The First Miracle Drugs: How the Sulfa Drugs Transformed Medicine* (New York: Oxford University Press, 2007).
9. Among the historiography of penicillin's early days, see Gladys Hobby, *Penicillin: Meeting the Challenge* (New Haven: Yale University Press, 1985) and Robert Bud, *Penicillin: Triumph and Tragedy* (Oxford: Oxford University Press, 2008). A review of the historiography can be found in María Jesús Santesmases and Christoph Gradmann, 'Circulation of antibiotics: An introduction', *Dynamis* 31(2) (2011): 293–303.
10. Antonio Cazorla Sánchez, *Fear and Progress: Ordinary Lives in Franco's Spain, 1939–1975* (Maldon, MA: John Wiley & Sons, 2009); Ángela Cenarro Lagunas, *La sonrisa de Falange: Auxilio Social en la guerra civil y en la posguerra* (Barcelona: Crítica, 2006); Conxita Mir, Carmen Agustí and Josep Gelonch, eds., *Pobreza, marginación, delincuencia y políticas sociales bajo el franquismo* (Lleida: Universitat de Lleida, 2014).
11. Joan Serralonga i Urquidi, 'El cuento de la regularización sanitaria y asistencial en el Régimen franquista. Una primera etapa convulsa, 1936–1944', *Historia Social* 59 (2007): 77–98.
12. Esteban Rodríguez Ocaña and Ferrán Martínez Navarro, *Salud pública en España: de la Edad Media al siglo XXI* (Granada: Escuela Andaluza de Salud Pública, 2008), 83–90. See also Esteban Rodríguez-Ocaña, 'The politics of public health in state-managed schemes of healthcare in Spain (1940–1990)', in *Images of Disease: Science, Public Policy and Health in*

Post-war Europe, eds. Ilana Löwy and John Krige (Luxembourg: Office for Official Publications of the European Communities, 2001), 187–210.

13. For a reconstruction of the malnutrition in Spain during the Civil War and the 1940s, see Isabel del Cura and Rafael Huertas, *Alimentación y enfermedad en tiempos de hambre. España, 1937–1947* (Madrid: CSIC, 2007). On the polio situation in Spain during the same period, see María Isabel Porras Gallo, Mariano Ayarzagüena Sanz, Jaime de las Heras Salord, María José Báguena Cervellera, eds., Rosa Ballester Añón, Inês Guerra Santos, Juan Antonio Rodríguez Sánchez and José Vicente Toledo Marhuenda, *El drama de la polio. Un problema social y familiar en la España franquista* (Madrid: Los libros de la Catarata, 2013). On typhus, see Isabel Jimenez Lucena, 'El tifus exantemático de la posguerra española (1939–1943). El uso de una enfermedad colectiva en la legitimización del Nuevo Estado', *Dynamis* 14 (1994): 185–98.
14. Trueta, *Fragments.*
15. The 'Introduction' to Simon Schaffer, Lissa Roberts, Kapil Raj and James Delbourgo, eds., *The Brokered World: Go-betweens and Global Intelligence, 1770–1820* (Sagamore Beach, MA: Science History Publications, 2009) has been inspiring regarding the concept of penicillin as an object involved in such networks.
16. 'Un nuevo remedio contra la infección luética', *ABC*, October 17, 1943, 35. My thanks to Esteban Rodríguez Ocaña for this reference.
17. Ramón Castejón Bolea, *Moral sexual y enfermedad: la medicina española frente al peligro venéreo (1868–1936)* (Granada-Alicante: Universidad de Granada-Instituto de Cultura Juan Gil-Albert, 2001); Ramón Castejón Bolea, 'Access to penicillin for the treatment of syphilis in Spain in the 1950's: the role of the international agencies', *European Social Science History Conference*, Valencia, March 30, 2016; Ramón Castejón Bolea and Esteban Rodríguez Ocaña, 'El control de las enfermedades venéreas en España durante la década de 1950: los inicios del Proyecto España 8 de la OMS', *Congreso de la Sociedad Española de Historia de la Medicina*, San Feliu de Guixols, Girona, June, 2017.
18. On Salvarsan in the Spanish newspapers during the 1910s, see José Luis Fresquet Febrer, 'El 606 en España. Una historia de seis meses contada por la prensa', *Anales de la Real Academia de Medicina de la Comunitat Valenciana* 12 (2011): 1–23.
19. On the history of infections, see Michael Worboys and Flurin Condrau, eds., *Tuberculosis Then and Now: Interdisciplinary Perspectives on a Post-modern Plague* (Montreal: McGill University Press, 2010). On pharmaceuticals, Vivian Quirke and Judy Slinn, eds., *Perspectives on Twentieth Century Pharmaceuticals* (Oxford: Peter Lang, 2010); Christian Bonah, Christophe Masutti, Anne Rasmussen and Jonathan Simon, eds.,

Harmonizing Drugs: Standards in Twentieth-century Pharmaceutical History (Paris: Glyphe, 2009); Christoph Gradmann and Jonathan Simon, eds., *Evaluating and Standardizing Therapeutic Agents, 1890–1950* (Basingstoke: Palgrave Macmillan, 2010); Andrea Tone and Elizabeth Siegel Watkins, eds., *Medicating Modern America: Prescription Drugs in History* (New York: New York University Press, 2007); Christian Bonah and Anne Rasmussen, eds., *Histoire et medicament aux 19e et 20e siècles* (Paris: Gliphe, 2005).

20. Lesch, *The First Miracle Drugs.*
21. Gwyn Macfarlane, *Alexander Fleming. The Man and the Myth* (Cambridge, MA: Harvard University Press, 1984); Hobby, *Meeting the Challenge*; Bud, *Triumph and Tragedy.*
22. Milton Wainwright and Harold T. Swan, 'C.G. Paine and the earliest surviving clinical records of Penicillin therapy', *Medical History* 30 (1986): 42–56.
23. Andrew Cunningham and Perry Williams, *The Laboratory Revolution in Medicine* (Cambridge: Cambridge University Press, 2002); Kevin Brown, *Penicillin Man: Alexander Fleming and the Antibiotic Revolution* (Gloucestershire: Sutton, 2004). On the early penicillin trials, see Wainwright and Swan, 'C.G. Paine'.
24. E. Chain, H. W. Florey, A. D. Gardner, N. G. Heatley, Margaret A. Jennings, Jean Orr-Ewing and A. G. Sanders, 'Penicillin as a chemotherapeutic agent', *The Lancet* 236 (1940): 226–8; E. Abraham, E. Chain, C. M. Fletcher, A. D. Gardner, N. G. Heatley, Margaret A. Jennings and H. W. Florey, 'Further observations on penicillin', *The Lancet*, 238 (1941): 177–89.
25. On the penicillin obtained during the Second World War at the microbiology laboratory in Delft see Marlene Burns, 'Scientific research in the Second World War. The case for Bacinol, Dutch penicillin', in *Scientific Research in World War II. What Scientists did in the War*, eds. Ad Maas and Hans Hooijmaijers (London and NewYork: Routledge, 2009), 44–61.
26. Hobby, *Meeting the Challenge*: This is the title of Chap. 8. The endocarditis episode is on 162.
27. Hobby, *Meeting the Challenge*, Chap. 5.
28. Hobby, *Meeting the Challenge*; John Parascandola, ed., *The History of Antibiotics: A Simposium* (Madison: American Institute of the History of Pharmacy, 1980); Harry Marks, *The Progress of Experiment: Science and Therapeutic Reform in the United States, 1900–1990* (Cambridge: Cambridge University Press, 1997); Mark Weatherall, *In Search of a Cure. A History of Pharmaceutical Discovery* (Oxford: Oxford University Press, 1990); John Parascandola, 'John Mahoney and the introduction of penicillin to treat syphilis', *Pharmacy in History* 43 (2001): 3–13.

29. Hobby, *Meeting the Challenge*, 284: Hobby dedicated a long endnote to explaining the particularities of the process for patenting penicillin-manufacturing methods in the USA, which were 'of prime concern to the British'.
30. Alfred N. Richards, 'Penicillin: A statement released by the committee of medical research', *JAMA* 122 (1943): 235.
31. Howard Florey and Hugh Cairns, *Investigation of War Wounds, Penicillin: A Preliminary Report to the War Office and the Medical Research Council on Investigations concerning the Use of Penicillin in War Wounds* (London: War Office, 1943); Hobby, *Penicillin: Meeting the Challenge*, 122–4.
32. On this episode see Bud, *Triumph and Tragedy.*
33. Hobby, *Meeting the Challenge*, 123–4.
34. Peter Neushul, 'Science, government and the mass production of penicillin', *Journal of the History of Medicine and Allied Sciences* 48 (1993): 371–95.
35. The Comité Nacional may have been created in 1943, according to Guillermo Olagüe de Ros, *Del uso de la retórica en el discurso científico: a propósito de los programas de trabajo de Fidel Fernández Martínez (1890–1942) y Eduardo Ortiz de Landázuri (1910–1985). Discurso leído en el acto de recepción pública* (Granada: Real Academia de Medicina y Cirugía de Granada, 2001), 33. The Comité was first mentioned in the Official State Bulletin in November 1944; see Ministerio de la Gobernación, 'Orden de 4 de noviembre de 1944 por la que se dan las normas para el uso de "penicilina"', *Boletín Oficial del Estado*, November 8, 1944.
36. Mariano Jiménez Casado, *Doctor Jiménez Díaz. Vida y obra* (Madrid: Fundación Conchita Rábago de Jiménez Diaz, 1993), 348–55.
37. E. Ortiz de Landázuri and J. Perianes, 'Normas generales de la terapéutica penicilínica', *Revista Clínica Española* 17(5) (1945): 361–72, 366–8. This paper describes the Comite Nacional starting to function in late 1944.
38. Marks, *The Progress of Experiment*, 106–7. On the US Chemotherapy Committee see David P. Adams, 'Wartime bureaucracy and penicillin allocation: The Committee on Chemotherapy and other agents', *Journal of the History of Medicine and Allied Sciences* 44 (1989): 196–217, and David P. Adams, *'The Greatest Good to the Greatest Number': Penicillin Rationing on the American Home Front, 1940–1945* (New York: Peter Lang, 1991).
39. The role of the black market in Spain's industry and economy during the first decades of Franco's regime has been analysed by historians of economics and industry; see Carlos Barciela, 'Crecimiento y cambio en la agricultura española desde la Guerra Civil', in Jordi Nadal, Albert Carreras and Carles Sudriá, eds., *La economía española en el siglo XX.*

Una perspectiva histórica (Barcelona: Ariel, 1987), 258–79, and Jordi Catalán, *La economía española y la segunda guerra mundial* (Barcelona: Ariel, 1995). For an example of the availability of penicillin on the black market, see Jesús Chamorro, *Recuerdos de una vida* (Oñate: Diletante, 2002), the memoir of a rural doctor in post-war Spain. I am grateful to Jon Arrizabalaga for calling my attention to this text.

40. On the Spanish public house in Madrid, see Casado, *Doctor Jiménez Díaz*, 350. The pub's owner was a patient of Jiménez Diaz.
41. José Álvarez-Sierra, *Lo que cura la Penicilina* (Madrid: Más allá, 1945).
42. Florencio Bustinza, 'El descubrimiento de la penicilina, momento estelar en la Historia de la Ciencia', *ABC*, November 5, 1944. On Bustinza's career see his *Currículum Vitae*, at the 'Expediente Florencio Bustinza 0687', Archivos de la Real Academia de Ciencias, Exactas, Físicas y Naturales, kept at its library, Madrid. He was elected a member of the Royal Academy of Sciences in 1943.
43. Bustinza *Curriculum Vitae*.
44. On the creation of the CSIC see M. J. Santesmases, *Entre Cajal y Ochoa: ciencias biomédicas en la España de Franco* (Madrid: CSIC, 2001); On the Junta de Relaciones Culturales, Lorenzo Delgado, 'Las relaciones culturales entre España y Estados Unidos, de la Guerra Mundial a los Pactos de 1953', *Cuadernos de historia contemporánea* 25 (2003): 35–59.
45. Florencio Bustinza, *Diez años de amistad con Sir Alexander Fleming* (Madrid: M.A.S., 1961): stay in US 39–41, stay in London, 55–72, daughter's infection and penicillin treatment, 72. On the NRRL, see Neushul, 'Science, government', 374–81, and references therein. On clinical trials of penicillin in the treatment of syphilis, Marks, *The Progress of Experiment*, Chap. 4.
46. Macfarlane, *Alexander Fleming*, 226.
47. Juan P. González, Jaime Suñer Pi and Francisco González, 'Obtención de la penicilina y otras experiencias', *Medicina Clínica* 2(6) (1944): 473–82; A. Esteve and A. Oriol Anguera, 'Aportaciones experimentales a la farmacognosia penicilínica', *Medicina Clínica* 3(2) (1944): 116–28.
48. A. Ballabriga Aguado, 'Algunas consideraciones acerca de la penicilina y otras micoínas', *Medicina Clínica* 4(2) (1945): 152–6; R. Kutz Echave, 'El empleo de la penicilina en la práctica médica', *Revista Clínica Española* 19(5) (1945): 340–9.
49. Ortiz de Landázuri and Peiranes, 'Normas generales', mentions many treatments in Spain.
50. Antonio Crous Serdá, 'Septicemia estafilocócica tratada y curada con penicilina', *Medicina Clínica* 4(5) (1945): 415–19.
51. Ortiz de Landázuri and Perianes, 'Normas generales', 368–9.

52. Many relatives, colleagues and friends have generously shared their memories of these cures with the author.
53. Fleming to Jiménez Quesada, reproduced in Mateo Jimenéz Quesada, *De Fleming a Marañón. Anécdotas de mi archivo* (Madrid: Minaza, 1972), 140. See also Mateo Jiménez Quesada, *El tratamiento de las infecciones de garganta, nariz, oídos y vías bajas respiratorias con nebulizadores de antibióticos* (Madrid: Faure, 1947).
54. Quesada, *De Fleming*, 132–5. Spanish university professors, including many clinicians, visited the USA from 1943 on.
55. No-Do Noticiario N° 130ª, June 25, 1945.
56. See the reconstruction of a plague by Italian historian of economics, Carlo Maria Cipolla, *Faith, Reason, and the Plague in Seventeenth-Century Tuscany* (New York: WW Norton & Company, 1981). A recent reflection on this issue, focused on the concept of global health, appears in a book review essay by Christoph Gradmann: 'Biomedical Sciences', *Studies in History and Philosophy of Biological and Biomedical Sciences* 30 (2017): 1–5.

CHAPTER 2

Fleming in Spain: The Hero, the Icon and the Politics of Public Acclaim

The public fame of penicillin and its representation during Alexander Fleming's visit will be analysed through the rich material I have found in his papers and reports published in the Spanish press. This will enable me to explore the resonances between public homage paid to the new drug and the Catholic basis of the dictatorship, including the extent to which the representation of public acclaim was part of the social practices promoted by the political regime. It will also facilitate analysis of the constructed public image of the drama of science and medicine, and its representation in the particular character of Fleming and the new drug. The (dis)connections between Fleming and the Spanish political authorities are issues that inspire reflection on the agency of penicillin in post-war international relationships. As well as the public acclaim of science and medicine—of scientific activity as a manufacturer of commodities for a better life—the ideology of progress in science and medicine was constructed through the misery of the post-war years, providing an almost religious adulation.

On May 26, 1948, Alexander Fleming arrived in Spain as the guest of Barcelona and Madrid's local governments. During his 19-day visit, Spanish newspapers consistently reinforced his status as a scientific hero. One source for this chapter is the notes taken by Fleming himself, who, despite receiving all the credit during his visit was only partly responsible for penicillin's success.[1] The others originate from newspapers and documentaries produced by No-Do, part of the dictatorship's policy of

M. J. Santesmases, *The Circulation of Penicillin in Spain*, Medicine and Biomedical Sciences in Modern History,
https://doi.org/10.1007/978-3-319-69718-5_2

mass media control and censorship, intended to create a popular culture supportive of the new regime.[2] The interplay of Alexander Fleming's diary and news reports show the limitations of official sources as reliable narratives of his stay.

I am therefore faced with reconstructing Fleming's visit to Spain, and its impact, through a historiographical lens, to provide a social and cultural reading of the narrative produced by Fleming and the press. At times the sources conflict, challenging each other's version, but for the majority they complement each other in a fascinating way. This synthesis suggests that penicillin had already taken on a life of its own and this had a direct effect on Fleming: no matter what he said or did, he was already a fervently admired hero, publicly associated with the magic healing penicillin provided. At the same time, the autonomy of penicillin as a public object, independent of the political regime of the country in which it circulated, counteracted state powers while appropriating the cultures of fervour associated with the dictatorship. That is, the enthusiasm for the dictatorship displayed in press narratives and by a part of the population was transferred to the foreign figures of Fleming and penicillin. The British bacteriologist and his drug shared the symbolic space of celebrations in a calendar of festivities carefully selected by the regime, intended to create cultural references to contribute to the new ideological basis of Spain.[3] By supporting the Catholic Church, Franco provided his military authority with a religious foundation. Falange Española, the fascist party that took part in the rebellion and supported Franco, was losing political influence following the Allied victory, but used its summoning power to gather crowds in support of the regime.[4]

Fleming's visit took place at a time when Franco's government was attempting to retrieve diplomatic relations with geographically and politically neighbouring countries; at this stage, France and the UK. By the end of the Second World War, during which Franco had explicitly supported the Axis and particularly Hitler while officially remaining neutral, the Franco regime had made a dramatic U-turn to 'show support for the principles adopted at the Bretton Woods Conference in New Hampshire (USA) and the United Nations declaration of February 1944'.[5] In 1945 the Cortes—an assembly with members appointed by Franco—approved local elections and a charter of civil rights, *Fuero de los Españoles*. The pretence of prospective democratisation did little to convince international opinion, and the General Assembly of the United Nations in February 1946 confirmed the exclusion of Spain. Isolated from the new legal order, Spain was able to maintain connections through bilateral commercial agreements. The UK, under an explicitly anti-Francoist Labour government, began negotiations to renew trade between the two countries, to provide British households with the fresh fruit and other diet-enriching materials they demanded. Spain, as a traditional exporter of such products, was exporting preserved fish, fruits and olive oil to France and the UK by 1945.[6]

These negotiations were kept secret, with the final agreement presented to the British parliament as 'essentially a technical readjustment' necessary to secure trade with Spain, and free of political implications. British economic interests during the long and arduous post-war recovery contributed to stabilising the dictatorship and positioning it within the new international order. According to historian of economics Fernando Guirao, the British government's position 'reveals the modest impact Spanish opponents to Franco had on Western policy-making'. The UK also helped the regime make similar connections with the rest of Western Europe.[7]

Fleming's visit coincided with both these negotiations between Spain and the UK, which would be completed in December of that year, and the start of diplomatic moves towards a military agreement with the USA.[8] By the late 1940s the UK, France, Germany and the USA were the main suppliers of imports to Spain, as they had been prior to the Civil War, while half of Spanish exports at the time were sent to European markets.[9] So perhaps the ceremonious tone evident in Fleming's diaries and the press was directed at this agreement as much as penicillin, the drug being dynamically identified with Fleming and the UK: both had provided healing action.

Despite the trade agreement there was no British ambassador in Madrid, but the Consul in Barcelona, Henry Hobson. Two different norms governed the spaces of diplomacy and trade, separated by a balance between needs and desires, and such norms and the barriers between these spaces were reconstructed during the post-war, both by the Allies and the dictatorship. Fleming's visit was a social representation of such a balance. Although the medical function of penicillin demands particular consideration, the drug also fulfilled a significant social and political role, embedded in the construction of contemporary cultures, as well as the public imagination of infections and their cures by new commodities, provided by the scientific progress and technical successes of the Allied victory.

Fleming was accompanied in Barcelona by Solly Flood, the director of the city's British Institute, Derek Traverse, and the clinician Dennis Morton, and while in Madrid by Walter Starkie, who had founded the British Institute to entertain and influence the highly educated Madrilenian bourgeois while promoting English culture and language.[10] For the month he visited Spain, Fleming acted as representative of both penicillin and the UK, and indeed promoter of his mother tongue. As his words were constantly translated by those clinicians who could speak English, the language of the post-war order shared in penicillin's fame.

Fleming was at that time professor of bacteriology at the University of London and working at St. Mary's Hospital Medical School, where he had trained in medicine and bacteriology with Almroth Wright. He had been awarded the Nobel Prize for Medicine in 1945, a time when this award did not confer the status and recognition it would later in efforts to stabilise scientific leadership of the Cold War atomic age.[11] By that time, as Robert Bud has stated, penicillin was fully embedded in the cultures of national propaganda. In 1942, while penicillin was still an experimental drug and clinically untested, the British newspaper *The Times*—closely followed by *The Evening Standard*—gave credit for the discovery of a new anti-infective drug to both London's St. Mary's Hospital and Alexander Fleming. UK and US magazines soon followed suit, and by 1944 Fleming was on the cover of the US weekly *Times*. War propaganda was an agent in the forging of penicillin's public association with Fleming and the UK, through newspaper proprietors, and a number of influential clinicians and industrialists, all interested in 'public perception' for various reasons: prestige and economic interests being among the most influential. In the USA, penicillin was launched on the public market following years of restricted availability while use on the front was prioritised. In 1945 a book was published, which after introducing the new drug by describing contributions made by early researchers such as Fleming and the Oxford team, focused on the industrial achievements of penicillin's manufacture. *Yellow Magic* by US popular writer J. D. Radcliff portrays an agenda that intertwined medical—scientific, at the laboratory bench—achievement and industrial efficiency.[12] Penicillin would become one of the bases upon which the irresistible empire of consumption was built.[13]

Fleming's wife Sarah, usually known as Sareen, accompanied him to Spain. Born in Kilian in the North–West of Ireland, Sarah Marion

McElroy had qualified as a nurse and with her twin sister Elizabeth created a highly regarded, private nursing home in London. She carried on working after they married in 1915, but when Fleming returned from the front in the First World War—which he spent at the casino in Boulogne, France, along with Almroth Wright—the two sisters sold the nursing home and devoted themselves to their families; Elizabeth had married Alexander Fleming's brother John. Sarah and Alexander had a son, Robert, born in 1924.[14]

When Alexander Fleming came to Spain, he did not seem to be aware of the persecution that prevailed over civil and military life, underpinning the public's fervent loyalty.[15] Fear was a social, collective emotion, the force of which was comparable to that of hunger, poverty and illness.[16] The hope of overcoming such a landscape of misery was condensing in the new drug and its human representative. In the Spanish social geography of the military dictatorship, food was rationed and every published word censored. Penicillin was still scarce and managed by government officials as a privilege. The acclaim surrounding Fleming cannot be observed independently from the public life of a society governed by a dictatorship. Although medical authorities in Barcelona could be perceived as challenging the centralist authority of Franco's government by leading the invitation to Fleming, thereby ensuring he flew into Spain's second city, his presence there followed government-imposed norms—from overstatements of his achievements to the celebration of his persona and the drug itself—and a public reception from a controlled population accustomed to showering authorities and religious figures with adulation. As a miracle drug, penicillin was adored as fervently as if it were a virginal saint, or a dreaded dictator.

Landing in Barcelona: Medical Authority and Public Acclaim

The invitation to Fleming originated from the Barcelona Hospital of Infectious Diseases, Hospital del Mar. The director of the hospital, cardiologist Luis Trias de Bes, was an influential clinician, familiar with the most prevalent infections of the post-war period: smallpox in 1940, and between 1941 and 1942, typhus. The Hospital del Mar, located near a seaport, had its origins in the sixteenth century as a quarantine site for those bringing infections from overseas, and in the late nineteenth century had been used as an isolation hospital during yellow fever and cholera epidemics. After the Spanish Civil War it became fully dedicated to the treatment of infectious diseases.

'Barcelona wants me to land in Spain there – out of Madrid'. These words were on the first page of Fleming's trip notes along with the many names of those who would be by the Flemings' side during their visit.[17] The welcoming party—including Luis Trias de Bes—demanded '[m]uch handshaking and photographing. Sareen presented with a huge bouquet'. The tone of Fleming's notes is already evident: everything he would receive during the three-week visit—presents, applause at public events, continual requests for autographs—would be annotated as 'too much' or 'embarrassing'. At the foot of the aircraft steps were numerous officials, delegates and deputies, including the director of Iberia airlines in Barcelona, and a multitude of clinicians and their wives; the only women in attendance were wives, referred by news reports solely in relation to their husbands. The Spanish newsreel shows the crowd, a large group of men and women parting to let the Flemings pass, Trias de Bes at Alexander Fleming's side passing his hat from one hand to the other, and smiling women alongside Sarah Fleming.

While in Barcelona, Derek Traverse, head of the city's British Council, would join Fleming and Trias de Bes at every ceremony, social encounter, walk and visit. Fleming travelled throughout Spain as the famous character penicillin had made him. The enthusiastic and flattering tone of newspaper reports and Spanish newsreels boosted and participated in his public fame. Pompous texts devoted to penicillin and its discoverer concealed the role that Howard Florey, Ernst Chain, Norman Heatley, Ethel Florey and many others had played in purifying and testing the antimicrobial drug. The Oxford group were noted fleetingly in newspapers, and Fleming briefly commented on the collective nature of the early tests.

The politics surrounding Fleming's public acclaim reflected the misery felt across a Europe slowly recovering from devastation, and a post-Civil War Spain characterised by terror and poverty, in which a handful of privileged people—government officials and those who had managed to keep hold of their wealth—experienced no shortages. Into such misery came the publicly celebrated cure by a miracle drug, represented in the emotion directed at Fleming. Despite occasionally expressing embarrassment for everything he received from the authorities, academics, clinicians and the public, Fleming paraded his fame and that of penicillin and tolerated his fervent reception at a time when the majority of penicillin demand was not covered; no penicillin was manufactured in Spain, and the amounts received by the Comité were inadequate, an issue not mentioned in any news report. On his first day in Barcelona, answering journalists' questions at the Ritz about the possibility of producing penicillin in Spain, he replied that it was a 'matter of money', and added '[t]oday America is the main manufacturer and exporter. Business is business there.'[18] This issue is never mentioned in his diary. Fleming was putting distance between his own achievements regarding penicillin and industrial manufacturing of the drug, from which he, although receiving full recognition for the drug's discovery, did not receive any direct economic benefit. As a drug, penicillin itself could not be patented, only the production method could, and indeed had been by Andrew Moyer, at the NRRL in Peoria.[19]

Trias de Bes directly boosted Fleming's public fame on the day of his arrival in Barcelona, publishing an article in the city's daily *La Vanguardia*, in which he described Fleming as an 'extraordinary benefactor of mankind' who was to be received with all the 'splendour and distinction [*señorío*] such an illustrious guest deserves'.[20] 'In no other countries I visited have I been so kindly received as here', Fleming declared after his arrival.[21] On reaching the Ritz there were 'more talks, more photographs and a broadcast from the bedroom'. After lunch the Flemings went 'for [a] drive in a car at our disposal: [we] just drive around'. When rested, he was presented with a programme for his stay in Barcelona by Trias. With so many reporters, translators and others around, taking photographs and asking questions, Fleming noted the next day he had 'met too many people to remember'. Reports of Fleming's visit were published alongside news of the arrival of meat and flour in Madrid, to be distributed in rations. Fleming travelled in an aura of celebration and wealth, protected from the misery surrounding him.

The next day Fleming visited the gothic cathedral; 'despite the crowds, the driver just kept his hands on the horn and we were driven right to the Cathedral steps'. A photograph on the front page of *La Vanguardia* shows him walking along the popular avenue Las Ramblas, where members of the public—apparently unprompted by anything but recognition of the feted scientist—are swarming around Fleming while a young woman presents him with a bouquet of carnations and roses.[22] The Flemings also joined 'everybody in best clothes' for the Corpus Christi procession, a widely and publicly celebrated holy day.[23] Including Fleming in the Catholic spaces and celebrations in Barcelona both reinforced the Catholic base of Franco's power and increased the gratitude of the masses by situating penicillin and its 'modest discoverer'—as the news reports phrased it—in the crowds, which opened to embrace him: 'when appeared cheering clapping – most embarrassing. (…) After the procession there was a renewal of the clapping and cheering, which followed us all the way back to the hotel.' People sang as they approached him, gave him presents and one woman fell to her knees: penicillin had saved her life. While Fleming wrote 'embarrassing', news reports stated that he was extremely moved. The hero had never been so accessible, a stark contrast with the drug he represented. On returning to their room the Flemings found 'an enormous wreath of flowers, an entirely new experience. Every seat is covered with them.'[24] Late in the afternoon, church bells tolled as the procession finished and military salutes were fired from the castle of Montjuic, marking the solemnity of Corpus Christi day while also exhibiting features of the dictatorship: the church and the military governed public life as a political domain. *Generalísimo Franco* had created three separate ministries for the army, navy and air force, all headed by loyal military personnel, as were many other ministries: the Minister of Industry and Trade, Minister of Public Works and Minister of Interior (Home Office) were all military men, while the Minister of Foreign Affairs and Minister of Education were staunch Catholics.

The next day brought more crowds, in a busy agenda apparently intended to introduce and reinforce political authority of Franco's governmental organisation and bureaucracy, the Flemings met the British Consulate, the mayor and the provincial governor. Fleming found the mayor—previously 'an orthopaedic surgeon in Zaragoza'—'very genial' and the Diputación's office 'one of the most ancient and beautiful in Barcelona'. They 'saw the University and ended in the student bar: sherry and hors d'oeuvres'. Back at the hotel a sculptor was waiting to create a bronze bust of the celebrated scientist, closely followed by a tailor—Trías's wife's uncle—to take measurements for two suits, 'one, a light weight, for Seville and the second, of a slightly heavier material, for Madrid'. At a party in the British Institute he again met 'too many people to remember' and the 'cheering crowds' were once more in attendance when the Flemings left. Back at the hotel the parents of a woman saved by penicillin 'presented Sareen with a sable stole'. At the end of the day Fleming concluded, 'Everywhere I have been so far, I have found everyone most demonstrative, a little too much for my own comfort.'

The trip remained both celebratory and Catholic the next day with a visit to Montserrat, a Benedictine abbey near Barcelona: although fully embedded in the practices and symbols of Francoism, the abbey would become a sanctuary for supporters of Catalonian nationalism and leftists as the power of the dictatorship declined. The one Benedictine monk who spoke English acted as translator, and Fleming shared lunch at the *refectorio* in silence but for one voice reading in Latin. In his welcoming speech the prior announced that the life of one of the elder monks had been saved from septicaemia by penicillin, then Fleming 'had to say a few words in reply – poor words but they seemed to like them'. Fleming presented the prior with a mounted culture of penicillin. 'He was overjoyed with it and together with a short description which I wrote for him, it is being put with the treasures of the monastery.'

After spending almost all day at the abbey, visiting the library, gardens and museum, their wives rejoined them—women were not allowed in the inner part of the abbey—to attend *Vísperas*, the sung evening prayer. While signing autograph books and planning dinner back in Barcelona, Fleming was notified that there was 'a minister just come to the dining room and that I should not meet him – a fascist'. Instead the Flemings went to a small restaurant across the street and 'had a simple good meal. Then the proprietor refused to take payment. I seem to be a hero here in Spain.' Despite his complacency regarding the homage and celebrations surrounding him, which represented not only the wonders of penicillin but also the dictatorship, meeting this fascist minister was apparently a line he would not cross.

Fleming felt the reverence, respect and acclaim he received were better suited to 'Winston Churchill or Princess Elizabeth'. Having sat another hour for the sculptor, Alexander and Sarah Fleming went to a museum, then to lunch with 'a large official party headed by the deputy mayor' who spoke 'good American'. Fleming attended a football match between Spain and Ireland. Driven in a 'procession of cars to the private entrance' then seated in what he felt corresponded to a Royal Box, Fleming recorded, 'we seem to have special privileges'. Also at the match was Antonio Girón, very likely the fascist official Fleming had avoided the night before. At that time, Girón was the only member of fascist party Falange Española in Franco's cabinet, as Minister of Labour. Franco had increased the Catholic presence in his cabinet—reducing the fascist contingent—by the end of the Second World War. The two factions had been in dispute since the dictatorship began; it would be the state religion that won the enduring victory.[25]

La Vanguardia reported that Fleming, having received a great ovation, was asked at half-time to say a few words to a full stadium that exuded 'great solemnity'. To end a very long day Fleming was invited to attend a bullfight alongside Barcelona's dignitaries.[26] A traditional feature of Spanish public social life, the *corrida* started at 23:15, unusually late, suggesting scheduling may have been specially organised for Fleming. Seated in one of the bullring's best positions, Fleming again received a tremendous ovation lasting for over three minutes, something a columnist reporting on the event had never previously witnessed.[27] Fleming was told the bullfighting was 'by no means up to standard', but one of the toreadors dedicated a bull's head to him, and at the 2:00 a.m. finish he left the arena to another great ovation.

After a week of social events, lunches, dinners and all the customary Spanish pastimes, Fleming finally gave his first lecture at noon on Monday, May 31, after attending an exhibition and visiting a shoe shop. The owner, having 'been saved by Pen after motor smash involving chest', wanted to reward Fleming, who chose a pair in brown crocodile skin.

At the Hospital de Nuestra Señora del Mar—dedicated to 'Our Lady' after the civil war, it returned to l'Hospital del Mar in the democratic era—for the treatment of infectious disease, 'the illustrious discoverer of Penicillin' gave a presentation on septic wounds. The scientist was back. In *La Vanguardia*'s front-page photo, the former man-about-town, seated at the optical microscope as if working in the laboratory, now wore the symbolic white coat.[28] Fleming officially inaugurated the conference hall in which he would present three lectures, the last on penicillin and its uses. The room was full. The audience, spilling out into the corridors and room next door, included many renowned Barcelonian clinicians, public health officials and faculty members from the University of Barcelona Medical School.[29] The talks were simultaneously translated by physician Adolfo Ley, a neurologist who spent two years at Illinois' Northwestern University, funded by the Rockefeller Foundation in the early 1940s.[30] Ley sat in a small chamber within the hall, and the translation was relayed through loudspeakers.

In his talk about septic wounds, Fleming paid homage to the Catalonian physician and bacteriologist he had met when first studying bacteriology, Jaume Ferrán, who died in 1929, and the traumatologist,

Josep Trueta, now exiled in the UK. Fleming praised Trueta's 'methods of encasing septic wounds of the limbs in plaster', which had 'saved an immense amount of suffering to the patients'. News reports noted the talk was followed by seemingly endless thunderous applause.

Fleming then went to lunch as the guest—he would find out later—of a pharmaceutical manufacturer. His host then took Fleming to his factory, but showed him nothing but films and photographs, what Fleming described as '[p]ure advertisement'. After an hour's nap—having eaten too much—the evening was dedicated to Trías, who had chosen that day to be admitted as a fellow of the Barcelona Academy of Medicine. Fleming found the dinner at Trías's home later 'cheerful'; there were 'no speeches but I had to sign all the menus with appropriate remarks'.

The next day, before his second lecture, Fleming was taken to the old hospital, a fifteenth-century gothic building, and the fourteenth-century gothic Marine Museum, one of the oldest and most beautiful medieval buildings in town. The lecture was on lysozyme, a protein from body fluids and hen egg white able to kill germs and inhibit growth, which Fleming had identified before penicillin.[31] Although it proved ineffective against infections, it is thought that lysozyme helped attune Fleming to the antimicrobial activity of biological substances.[32] Lunch was with medical authorities, and dinner with local government officials. This 'full dressed' event was held at the Palacio de la Diputación—today Palau de la Generalitat—a fifteenth-century building built for the Catalonian government, and one of the few medieval buildings in Europe to still house the institution it was originally constructed for. By now Fleming was overwhelmed by the amount of autographs he had to provide: a diverse array of notebooks, menus and pieces of paper awaited his signature.

Finally, Fleming presented his lecture on the uses of penicillin. This was a talk he had given many times, with an added discussion on treatment with penicillin and the anaesthetic procaine.[33] Initially injected separately to reduce pain—the acids would destroy the stomach if taken orally—these chemically combined as a procaine salt of penicillin was one of the earliest successful combinations of penicillin.[34] After the talk, medical staff invited Fleming to an 'enormous' lunch at a popular restaurant close to both the hospital and docks, which included 'fish and suckling pig'. He was 'completely defeated but the rest ate a lot, together with sherry, red and white wine, champagne and brandy'; rationing allowances were apparently not in effect. An afternoon cocktail party at the British Club, where he was introduced to 'over 100' people, all wanting autographs, was followed by a concert at the Palau de la Música. The Flemings left before the end, but being unable to slip away discretely they were treated to lengthy applause, a large box of chocolates and an 'enormous bouquet of flowers'.[35]

The next day—free from lectures but still with a full agenda—the Flemings were taken to the beautiful seaport and beach at nearby Sitges, a trip described by Fleming as 'flowers, terraces, ovations, ceramics, autographs, dancing'. In the afternoon another grateful recipient of penicillin awaited, this time an optician, who made him two pairs of glasses, for both of which Fleming chose identical frames to those he was already wearing. At the Academy of Medicine that evening the tuxedoed Fleming gave a four-minute talk on the history of penicillin, and ended by stating his desire to 'strengthen the cultural, friendly relationship between my home country and Spain'. Trías then read a complete translation.[36] After a 'flowery' speech by the president, Fleming received a diploma and a gold medal coined for the occasion. A No-Do film of the event—the hall had to be highly illuminated for this—and *La Vanguardia*'s cover, show Fleming in the full *Aula magna* of the Academy, holding the diploma and bowing in acknowledgement of the ovation he received.[37] Applause accompanied the Flemings as they walked to a car provided by the British consulate. As *La Vanguardia* reported, '[t]he wise man kindly waved at the crowd'.[38]

Fleming visited the Hospital Santa Cruz y San Pablo's campus the next day, a set of early twentieth-century modernist buildings and the first large hospital in Barcelona, where he noted religious regalia—'Altars in all the wards'—alongside 'modern equipment'. In the maternity wing he witnessed the conditions within the unmarried mothers' hall: an 'enormous place', with '400 beds' where the women were 'visited by no-one but the physicians ... The children, if necessary, are cared for here until they reach 6 years of age.' In the afternoon Fleming was expected to present a talk on 'Success' at the University of Barcelona. Such a sociological issue by a renowned researcher, 'Success' was a direct promotion of not only his own achievements as a scientist, but of research as such, of bacteriology and the clinical laboratory; the prestige of St. Mary's Hospital was included. The University of Barcelona Medical School's large auditorium—Paraninfo—was full of 'enthusiastic' students and a Spanish translation was provided. 'Judging by the ovation', Fleming noted, 'it was apparently successful',[39] while *La Vanguardia* reported extended applause for 'the British sage'.[40]

Back at the Hospital del Mar on Saturday, Fleming attended a 'great'—and apparently boisterous—reception: '[a]bounding with policemen to keep order'. This time the new research department of tropical diseases was official inaugurated. After a number of speeches by hospital dignitaries, including Trías, Fleming spoke on what he believed a research building should be. After admiring the new Bacteriological Institute, he added that 'a building is not everything' and proceeded to praise the modesty of the research profession: 'We laboratory workers do not wish marble halls and majesty but we like to look on the institute in which we work with a certain amount of pride.' Such a 'magnificent building' should be justified, he wrote, 'by the high quality of their work', and he ended with a final remark: 'I do not doubt that in this they will succeed and that the city of Barcelona will continue to be as proud of this bacteriological institute as I am, this day, proud at being allowed to assist at its opening.'[41] The officials, clinicians and Fleming then proceeded to the entrance hall, where a curtain was pulled back, and 'there appeared my head in clay', alongside a commemorative plaque of his inauguration of the Institute.[42] Following champagne, they visited the laboratories, which Fleming thought 'much too small. There were only single gas jets, and very little in the way of equipment.' It was during this decade that a number of research institutes in Madrid were also opened only to remain empty of research material and personnel: the dictatorship expressed its power through impressive buildings without providing funds to sustain the work for which they had been constructed.[43]

Social engagements continued, including a visit to the emblematic Pueblo español, a presentation of buildings from the different regions of Spain created for the 1929 Barcelona International Fair. On Sunday, Fleming was back at Las Ramblas for the last few crowded walks, ovations and demands for autographs that maintained the tension and acclaim. He also viewed the early version of a newsreel with images of his stay, closing the circle: watching himself perform as the greatly acclaimed penicillin hero. There remained a final bullfight—'the whole 20,000 people applauded so have to wave my hand in all directions'—then packing, made harder by the number of presents and other souvenirs the Flemings had accumulated.

On Monday morning, the Flemings boarded a plane, alongside Trías and his family, and waved the Barcelonian crowds a final goodbye. They landed in Madrid, and 'after an hour of extraordinary fuss' flew on to Seville where they found an 'even greater reception, many doctors, more flowers'. The people of Seville had been notified of the Flemings' visit to the town by their mayor, who called for them to honour the British hero, 'to render this great benefactor of mankind the homage we are all compelled to pay to him'.[44] This call—in tune with the tendency of authorities to summon people to attend appearances and public speeches by Franco—would be answered in abundance. It was early June in one of Spain's hottest cities, during a drought that had blighted the entire decade and extended post-war shortages of food and water, when the crowds came to see Fleming arrive at Seville's town hall. The city, her citizens, journalists and officials—if we take the information in newspapers to be reliable—exhibited even higher, more pompous overtones than Barcelona. Fleming was practically venerated, afforded an excess of respect normally reserved for God, the Holy Virgin and, whether through actual respect or fear, the Dictator himself. Power, wisdom, miraculous wonders and desperate expectations, all coalesced in the British hero.

News reports described Doctor—the same Spanish word is used for MD and PhD—Fleming's outfit in detail, as well as Sarah's, as they visited Seville's principal buildings—the cathedral and the old Hospital de la Caridad—before meeting the crowds in the town hall square, Plaza Nueva. Local authorities—the bishop, the army, police and clinician members of the Academy of Medicine—were all represented, as the mayor made a speech in Spanish, 'with no translation', then Fleming 'had to make a short one'. The Municipal Band played the British national anthem as Fleming passed by and went up the stairs to the hall where he found a group of people cured by penicillin and thunderous applause: news reports may have exaggerated but Fleming's notes suggest his reception, here and everywhere else, was enthusiastic, including 'more applause in the street'. Dinner at the mayor's was accompanied by a group of young girls performing Andalusian dances, *flamenco* singing and many more autographs to sign. The Flemings managed to get to their hotel by 2:30 a.m.: 'Not bad for a free afternoon.' After paying respect to the great hero in the chronicles of his first day in Seville, the local edition of *ABC* reproached Fleming for always being late: 'We as polite Sevillians and respectful of British punctuality, are always on time for every event. And Doctor Fleming politely corresponds to our idiosyncrasy by always being late.'[45]

The theatre for Fleming's lecture was decorated with Spanish and UK flags, flowers and a long red carpet along the aisle from the stage.

A large portrait of Franco dominated the scene, but went unmentioned by Fleming.[46] His lecture on the history of penicillin was read in Spanish by surgeon Cristóbal Pera. A visit to downtown Seville followed, with 'Alcázar [a royal palace], patios, narrow streets in daylight': Fleming was apparently charmed.

The Flemings and the Trías de Bes family were driven to Córdoba the next day, through virtually continuous olive groves, to be officially greeted at the main door of the city, a Roman bridge over the river Guadalquivir. Fleming himself had requested the visit to Córdoba be arranged.[47] They walked to the mosque—through 'an enormous crowd, great cheering and much handshaking'—which impressed Fleming: 'A marvellous building quite different from anything I had ever seen.' He visited the museum of painter Romero de Torres, attended another town hall reception as guest of honour, and had lunch at the Club de la Amistad, described by Fleming as 'the most aristocratic club in Cordoba'. The patio of the restaurant was hung with Union Jacks and Spanish flags, and Fleming was sat between the governor's wife, who only spoke Spanish, and the bishop, who had a modicum of English: 'Sareen was better off since one of her neighbours spoke English.' He was introduced to the first women in Cordoba to receive penicillin, and was given a *Cordobés* hat as a gift, with which he posed for photographers. After visiting a bullfighting exhibition at the Veterinary School—'Evidently bullfighting has [a] great part in Spanish life'[48]—the Flemings and the Trías de Bes family returned to Seville (see Fig. 2.1).

After overnighting in Seville, the group travelled south to Jerez de la Frontera for lunch in one of the wine cellars, with a number of producers and the Mayor of Jerez: the town is the origin of the British term, sherry. After a display of 'very beautiful' Andalusian horses at the Spanish Military Stud, they returned to Seville.[49] The Flemings flew to Madrid the next day for a long, intense weekend.

At the Madrid aerodrome representatives of the Ministry of Education, local officials, representatives of the University of Madrid's School of Medicine, director of the British Institute in Madrid, Walter Starkie, and Florencio Bustinza and his family, were all waiting.[50] Fleming received a letter of gratitude from the aerodrome porter for penicillin saving his life after surgery and was driven to the city with the major and Starkie. From a downtown broadcasting studio, Fleming greeted the people of Madrid.[51] His first talk was the one on 'Success' presented at the British Institute, and introduced by Bustinza who, as

Fig. 2.1 Fleming in Córdoba before Manolete statue at the bullfighting exhibition, Veterinary School in Córdoba, shaking hands with the statue's creator Laiz Campos. Photographer unknown. Courtesy of Margarita Laiz

discussed in the previous chapter, had written a book on penicillin after a long stay in the USA and a visit to Fleming. It was during Fleming's stay in Madrid that Bustinza displayed his knowledge and social skills. The Institute was full: 'there were perhaps 200 people listening but I did not know that there were 3 or 4 times in different floors'. That evening Fleming went for dinner with the Starkies, Trias and Bustinza at the exclusive country club, Club de Campo, where he met the Duke of Alba, Jacobo Fitz-James Stuart, who had been Spanish Ambassador in London, then attended a dance in aid of the Anglo-American hospital; although they did not stay long, Fleming was pleased to hear their appearance had attracted a large attendance and brought in a substantial amount of money.

In Madrid Fleming received all the accolades that existed for the most illustrious dignitaries. The Minister of National Education, José Ibáñez Martín, informed Fleming that Franco had decided to award him the highest honour for those who excelled in education and culture, the *Gran Cruz de Alfonso X el Sabio*; Madrid's local government named a street after him; and the next day, Saturday June 12, in a ceremony at the University of Madrid at which he received the *Gran Cruz*, he was also invested as Doctor *honoris causa*. A photograph in Madrid's daily *ABC* shows Fleming receiving the doctorate from the University Rector, with José Ibáñez Martín between the two looking directly into the camera. The photograph shared the page of the ceremony report with a caricature profile of Fleming in a satin biretta and cape.[52] His lecture was once again introduced by Bustinza, who also read the Spanish translation.[53] On leaving for his hotel after lunch, people ran alongside the car from the *rectorado* all the way down San Bernardo Street. Fleming presented his lecture on penicillin use at a downtown theatre late in the afternoon, to an audience mainly composed of clinicians and pharmacists.

On Sunday, Fleming visited the Prado museum, then went for lunch in Toledo, at the country house of the well-respected clinician, Gregorio Marañón: 'A very nice family and a beautiful house. We posed for many photographs and cine pictures. We lunched outdoors and it was all most pleasant.' The entire Marañón family was there to receive Fleming, the Starkies, the Trias family and Bustinzas; Sarah Fleming was in bed with a fever that weekend, being cared for by a nurse from the Anglo-American hospital. In the afternoon Fleming received honorary membership of the Royal Academy of Medicine. Many academic clinicians and José Ibáñez Martín were again in attendance: Fleming does not document any meetings with authorities and the many other influential people he encountered, as if barely a word was spoken between them. After a long address by the Academy's President, Fleming, wearing his 'new decoration' for the first time, gave a far briefer speech. He was introduced to the Bishop of Madrid, a representative from the Ministry of Foreign Affairs, and Fernando Enríquez de Salamanca, Director of the Academy and Dean of the University of Madrid Medical School; de Salamanca signed the reports which led to the purging of members of the Faculty judged to be disloyal to Franco during the early 1940s. That Sunday evening in Madrid, Fleming was surrounded by representatives of power: the church, government, medicine and academia. Nowhere more dramatically than Madrid was Fleming so close, face to face with the dictatorship. As a fully centralised government, the capital city of the nation remained the most concentrated representative scenario of the powers of Franco's regime. After the academic session Fleming visited Chicote, which he described as a 'drink Museum', but was actually a restaurant where the owner, Perico Chicote, kept a collection of drinks from many countries, including 222 different brands of Scotch whisky: 'I left him with 221.' Among the guests at this 'very high class cocktail party', were 'film stars, diplomats, duchesses, doctors and *The Times* correspondent and I do not know who else'. Chicote was one of the most recognised places in Madrid to illicitly purchase penicillin: Fleming's presence would have been beneficial for both the cocktail bar and the sales of black market penicillin. Next was a musical soiree at the Parque del Buen Retiro in formal dress, with the mayor and over a hundred guests: Fleming again received a great ovation, shook many hands and signed countless autographs.

On Monday afternoon, after broadcasting a farewell speech,[54] the Flemings left for the airport 'in two cars, with a lorry for the baggage': Bustinza noted that a new suitcase had to be acquired to carry the numerous presents Fleming had received.[55] After a 45-minute wait at Bordeaux they arrived at RAF Northholt just before 9:00 p.m., 'having flown over a half wrecked Mulberry Harbour'. These temporary portable harbours developed by the British during the Second World War facilitated the offloading of cargo during the Allied invasion of Normandy in June 1944. Fleming's typewritten report ends here, remembering British participation in the war.[56]

A Symbol in the Dictatorship's Landscape

Fleming has left his own story of the lengthy and acclaimed visit he made to the impoverished country that provided Britain and France with oranges. At the time of his visit penicillin supplies in Spain remained inadequate to meet the population's considerable need. The old governing aristocracy had been replaced by a new order, one which took full advantage of its links with Francoist bureaucracy, resources that enabled rationing regulations to be flouted and benefits to be accrued from the inflated price of agricultural products on the black market.[57] Fleming's availability during his visit—allowing people to see him, touch him, receive his autograph—counteracted the misery of everyday life in Franco's Spain. He was penicillin's representative in a culture of public acclaim: the culture Spanish citizens had become accustomed to. The mythic mobilising discourse of Franco's Spain, an imagined community protecting the country from Republican ideals with historic, traditional values, were narrative tools put at the service of penicillin as a myth and its representative, Fleming.[58] Penicillin and the dictator's imposed ideology mutually reinforced the status of each as salvation. By applauding public authorities—the raised-arm fascist salute had not been compulsory since 1945—Spanish citizens became part of a public policy and culture of acclaim ordered and showcased by the dictatorship. Fleming's fame and the mythic status of penicillin originated from a cure experienced by only a handful of people. As public celebrations, the events which Fleming attended could be regarded as acknowledgements of a well-deserved fame, of miracles and medical and scientific authority. Fleming never met Franco, and the dictator does not appear in his notes, as if non-existent. Apparently Franco did mention Fleming, however, when authorising the *Gran Cruz de Alfonso X el Sabio*: the dictator recognised Fleming.

In addition to visiting academic settings, new and established hospitals, and local and regional government institutions, Fleming attended bullfights, football matches, concerts and processions. He also visited restaurants, private homes, churches, museums and wine cellars, and walked down Spanish streets. Joining in contemporary features of Spanish public social life with government officials and distinguished clinicians from Barcelona, Seville, Córdoba and Madrid, Fleming ostensibly shared the values of that space and time.[59] Apart from the occasional embarrassed smile, Fleming's thoughts on the political and social situation in Spain—if indeed he had any—remained absent from both his speeches and his journal.

Honoured by the Academies of Medicine in Barcelona, Seville and Madrid, he gave his presentations on microorganisms, antimicrobials and success again and again; the audiences were extensive, listening next door to translated versions, reading news reports, watching No-Do reels or running beside the cars in which he and Sarah Fleming were driven. Throughout his tour he was publicly acclaimed as the heroic discoverer of a wonder drug.[60] The English language was present, participating in a new phenomenon of propagating modern geopolitics through the idiom of the penicillin makers: the language of those who worked in bacteriological laboratories, and in the British and American manufacturing plants.

In Madrid he was proclaimed a 'peace messenger of medical science', thanked for his 'example and simple life as a laboratory worker', and for his work, which 'benefited everyone equally'.[61] The record of these events, detailed in newspapers, recorded by No-Do and published by Fleming's Spanish spokesman, Florencio Bustinza, carries the public tone of the period: a heroic individual celebrated at a time when groups were considered a threat to civil life by a highly authoritarian and controlling dictatorship. Public imagery included both the positive and dark sides of penicillin: Fleming even visited one of the famous public houses in Madrid where penicillin was sold *de estraperlo*, a popular term for the black market, on which many basic foods such as bread, sugar and coffee were also traded and sold. Meals were lavish feasts at which he always overindulged, while Spanish citizens had to adhere to strict rationing regulations. Busts of the legendary, mythical saviour appeared across Spain following his visit, and Fleming's fame would increase over the years, even when penicillin was no longer the barely accessible, magic drug.

Finally, beyond the hero stood the representative of Britain, aiding the recovery of relations between Spain and the UK. Trade and diplomacy were inextricably linked and the British Consulate was a constant component of Fleming's entourage. As Fleming noted, the Consul General was 'very pleased I have come as I will do a great deal to help relations between Spain and our country'. Although his biographers have emphasised Fleming's modest character, he recognised his diplomatic value: 'It seems to me that I am much more important as an ambassador than as a medical figure' (56118 5). This was certainly not Fleming's first trip abroad. His journey to the USA in 1945 to witness 'how penicillin was being developed there', had also become a propaganda mission, with talks, broadcasts, interviews and press conferences. On arrival Fleming discovered that British representatives believed 'the most important aspect of my visit would be in respect of international amity and that it was very important that I should not discourage publicity'.[62] Andre Maurois's biography of Fleming describes his stay in Spain as 'une voyage des Mille et une Nuits', exhausting, demanding, and without time to rest.[63] He and his wonder drug shared their promise of salvation with the dictatorship through an interchange of symbols, celebrations and styles of myth-making that suggested overlapping values. The heroic dictator thus interchanged ways of celebrating with penicillin itself and with Fleming, in a set of events that if not produced by, were at least permitted by political authorities who rigorously controlled every public—and many times private—expression.[64]

The media played their part in creating the myth that penicillin and Fleming became. Newspapers published in-depth reports of each day of Fleming's visit and announced plans for the next. The detailed account this chapter provides cannot do justice to the overtones, the honours—medals, diplomas, street names, the *Gran Cruz* decoration—or the mobbed welcomes and walks reported by No-Do and the papers. The extremely tight schedule added to the mythical journey, the circulation of a scientific hero in the country of the military hero, at a time when the successes of science and technology were considered a lifeline for the future. All these events would be obscured by approval of the Marshall Plan budget to aid European recovery: Spain was omitted from the countries to benefit. Spain relied on scientific achievements arriving from abroad, but it was an apparently wealthy nation displayed to Fleming.

The bullfight is a particularly apt symbolic space for the story reconstructed in this chapter. The dangers could be perceived as surmountable, possible to overcome with the contribution of penicillin, of the science and medicine of the new drug. Penicillin was saving the lives of those thrust from the bull's horns, whose wounds became infected: many bullfighters died before penicillin was available. In Córdoba Fleming had seen the monument to the bullfighter and 'national hero' Manolete, killed the previous year. The myth would survive, however, even when the manufacture of penicillin in Spain had begun.

Notes

1. All Fleming's quotations included here are from Alexander Fleming Papers, British Library Additional Manuscripts, BL Add MS 56191, Vol. LXXXVI Journals, 25 May–14 June 1948, handwritten notebooks, and from the typewritten version of the journals in BL Add MS 56118. I am indebted to Mauro Capocci for calling my attention to this archival material.
2. F. Sevillano Calero, *Propaganda y medios de comunicación en el franquismo* (Alicante: Universidad de Alicante, 1998), at http://publicaciones.ua.es/filespubli/pdf/LD84790874637062078.pdf.
3. Ángela Cenarro, 'Los días de la "Nueva España": entre la "revolución nacional" y el peso de la tradición', *Ayer* 51 (2003): 115–34.
4. Ismael Saz, *España contra España: Los nacionalismos franquistas* (Madrid: Marcial Pons, 2003); Ismael Saz, *Las caras del franquismo* (Granada: Comares, 2013).
5. Jefatura del Estado: Decreto-Ley de 5 de mayo de 1945, published by the *Boletín Oficial del Estado*, 8 de mayo 1945: 'Bloqueo de bienes extranjeros'. On this episode see also Jean-Marc Delaunay, 'La liquidation des avoirs allemands en Espagne (1945–1961)', in *España, Francia y la Comunidad Europea*, eds. Jean-Pierre Étienvre and José Ramón Urquijo (Madrid: CSIC-Casa de Velázquez, 1989), 219–45.
6. Enrique Moradiellos, *La España de Franco, 1939–1975: política y sociedad* (Madrid: Síntesis, 2000); Fernando Guirao, *Spain and the Reconstruction of Western Europe, 1945–57. Challenge and Response* (London and New York: Macmillan, St. Martin's Press, 1998).
7. Guirao, *Spain and the Reconstruction of Western Europe*, 28–34.
8. On the military agreement between Spain and the USA, see Ángel Viñas, *Los pactos secretos de Franco con Estados Unidos: bases, ayuda económica, recortes de soberanía* (Barcelona: Grijalbo, 1981).
9. Guirao, *Spain and the Reconstruction of Western Europe*, 28–34.
10. On Starkie, see Isabel Bugallal, 'Walter Starkie, el Borrow irlandés', at http://www.laopinioncoruna.es/contraportada/2010/12/14/walter-starkie-borrow-irlandes/447969.html; Jacqueline Hurtley, *Walter Starkie. An Odyssey* (Dublin: Four Courts Press, 2013).
11. Gwyn Macfarlane, *Alexander Fleming, the Man and the Myth* (Cambridge, MA: Harvard University Press, 1984); Kevin Brown, *Penicillin Man: Alexander Fleming and the Antibiotic Revolution* (Stroud: The History Press, 2013); and André Maurois, *La vie de Sir Alexander Fleming* (New York: EP Dutton, 1959).
12. Bud, *Triumph and Tragedy*, 63–4.

13. Inspired by Victoria de Grazia, *Irresistible Empire: America's Advance through Twentieth-Century Europe* (Cambridge, MA: Harvard University Press, 2005); John Druty Radcliff, *Yellow Magic. The Story of Penicillin* (New York: Random House, 1945), cited in Bud, *Triumph and Tragedy*, 67.
14. Brown, *Penicillin Man*, 66–8.
15. Among the many histories of the Franco dictatorship, see Moradiellos, *La España de Franco*; Ismael Saz, 'El primer franquismo', *Ayer* 36 (1999): 201–21.
16. Antonio Cazorla Sánchez, *Fear and Progress: Ordinary Lives in Franco's Spain, 1939–1975* (Maldon, MA: Wiley, 2009). Also Saz 'El primer franquismo'.
17. One of Fleming's biographers had suggested it was Bustinza's arrangements that made Fleming's tour of Spain a 'triumphal visit': Gwyn Macfarlane, *Alexander Fleming*, 227. However, Bustinza's reconstruction of the trip and reviews in newspapers suggest Trías de Bes was the principal instigator. Bustinza did make academic arrangements for Fleming in Madrid: Florencio Bustinza, *Diez años de amistad con Sir Alexander Fleming* (Madrid: M.A.S., 1961). On Fleming's similarly acclaimed visit to France in 1945, see Jean-Paul Gaudillière, *Inventer la biomedicine: la France, l'Amerique et la production des savoirs du vivant* (Paris: La découvert, 2002), 46–7.
18. *La Vanguardia*, May 27, 1948, 9. My translation.
19. On penicillin patents see Gladys Hobby, *Penicillin: Meeting the Challenge* (New Haven: Yale University Press, 1985), 284, no. 18; Ana Romero de Pablos, 'Regulation and the circulation of knowledge: Penicillin Patents in Spain', *Dynamis* 31(2) (2011): 363–83; and Ana Romero de Pablos, 'Patents, antibiotics, and autarky in Spain', *Medicina nei secoli* 26(2) (2013): 423–49.
20. Luis Trias de Bes, 'Alexander Fleming y su penicilina', *La Vanguardia*, May 26, 1948. This was reproduced and commemorated by *La Vanguardia digital*, August 23, 2002, as one of the 'Articles of the century', at http://www.lavanguardia.es, accessed 24 October 2002.
21. *La Vanguardia*, May 27, 1948, 9.
22. *La Vanguardia*, May 28, 1948, cover page.
23. Javier Tusell, *Franco y los católicos: la política interior española entre 1945 y 1957* (Madrid: Alianza, 1984).
24. Fleming Papers 56118.
25. Moradiellos, *La España de Franco.*
26. *La Vanguardia*, June 1, 1948, 13.
27. E. P., 'Fiesta de los Toros', *La Vanguardia*, June 1, 1948.

28. *La Vanguardia*, June 2, 1948. Similar pictures were also included in other newspapers around that time, such as the Barcelona weekly, *Destino*, May 22, 1948.
29. *La Vanguardia*, June 1, 1948, 8.
30. See his speech on entering the Academia de Medicina de Barcelona, the response and biographical information about Ley in Adolfo Ley Gracia, *Pasado, presente y future de la cirugía craneocerebral.* Discurso leído por el académico electo en el acto de su recepción (Barcelona: Real Academia de Medicina de Barcelona, 1971).
31. *La Vanguardia*, June 2, 1948, 7.
32. Hobby, *Meeting the Challenge*, 7–8 and references therein.
33. The lectures were collected and published in Barcelona later that year, with a foreword by Trias de Bes: Alexander Fleming, *Conferencias magistrales* (Barcelona–Madrid: Publicaciones del Hospital Municipal de Infecciosos-Diana Artes gráficas, 1948).
34. Bud, *Triumph and Tragedy*, 62 and references therein.
35. *La Vanguardia*, June 4, 1948, 7.
36. *La Vanguardia*, June 4, 1948, 7.
37. *No-Do* 283A, año V; *La Vanguardia*, June 5, 1948.
38. *La Vanguardia*, June 4, 1948, 7.
39. Second notebook of the visit '4–14 June 1948'. Fleming Papers 56192.
40. *La Vanguardia*, June 5, 1948, 7.
41. Speech Opening Research Institute Barcelona 1948, British Library, Fleming Papers 56125.
42. *No-Do* 283A, año V.
43. Maria Jesús Santesmases, *Entre Cajal y Ochoa: ciencias biomédicas en la España de Franco* (Madrid: CSIC, 2001).
44. *ABC Sevilla*, 'Mañana llegará Mr Fleming a Sevilla' [Tomorrow Fleming arrives in Seville], June 6, 1948, 11.
45. Gil Gómez Batuelo, 'El señor Fleming, humorista', *ABC Sevilla*, June 9, 1948, 5.
46. *ABC Sevilla*, June 9, 1948, 5.
47. *ABC Sevilla*, June 10, 1948, 6.
48. Some details of the daytrip are also in *ABC Sevilla*, June 10, 1948, 6.
49. *ABC Sevilla*, June 11, 1948, 4; *ABC Madrid*, June 11, 1948, 11.
50. On Starkie, see Bugallal, 'Walter Starkie'; and Hurtley, *Walter Starkie.*
51. *ABC*, June 12, 1948, 7.
52. *ABC*, June 13, 1948, 15.
53. Bustinza, *Diez años*, 88–100.
54. *ABC*, June 15, 1948, 17.
55. Bustinza, *Diez años.*
56. Fleming Papers 56118.

57. Miguel Artola Blanco, *El fin de la clase ociosa: de Romanones al estraperlo, 1900–1950* (Alianza Editorial, 2015).
58. Benedict Anderson, *Imagined Communities: Reflections on the Origin and Spread of Nationalism* (London: Verso Books, 2006). I was directed to Anderson by Zira Box Varela, *España, año cero: la construcción simbólica del franquismo* (Madrid: Alianza Editorial, 2010).
59. *La Vanguardia*, June 1, 1948, 8.
60. A short chronicle of his trip to Sevilla, Córdoba, Jerez and Madrid is in Bustinza, *Diez años*, 86–7. On Fleming's stay in Madrid see also Mateo Jimenéz Quesada, *De Fleming a Marañón. Anécdotas de mi archivo* (Madrid: Minaza, 1972), 141–2.
61. *ABC*, June 12, 1948, 7.
62. Brown, *Penicillin Man*, 173.
63. Maurois, *La vie de Sir Alexander Fleming*, 255.
64. On the symbolic construction of Franco regime, see Box, *España año cero*, 43.

CHAPTER 3

Manufacturing Penicillin: Industrial Policy, Gender and the Antibiotic Factory

A detailed reconstruction of early negotiations on the building of two penicillin factories in Spain is combined in this chapter with a gender approach to the organisation of factory work. This enables me to explore industrial and health policies of the 1940s, and their relation to their recent past: public health and medical practice prior to the Civil War and the Second World War. Links between these two professional and political domains will be analysed to account for the negotiations with US firms that led to the production of penicillin, and later other antibiotics, by two government-sanctioned firms. Protection and control strategies combined with foreign industrial and scientific relations made industrial agreements possible, initiating domestic penicillin manufacture from 1950 onward. The participation of women in these factories and the division of labour between men and women will be explored in relation to the symbolic gendered values associated with penicillin.

Once the Comité Nacional took up its role as distributer of the batches of US penicillin, the need to make judicious use of such limited supplies became a central aspect of its introduction to Spain.[1] A 1944 report by a member of the Comité stressed the economic importance of using it in the smallest amounts possible.[2]

In 1945, a full report on methods of extraction and clinical trials of a pharmaceutical product from *Penicillium notatum* was submitted to the Pharmaceutical Products Section of the Dirección General de Sanidad (Directorate-General for Health). This was not a pure drug, but rather

M. J. Santesmases, *The Circulation of Penicillin in Spain*, Medicine and Biomedical Sciences in Modern History, https://doi.org/10.1007/978-3-319-69718-5_3

the concentrated extract from a mould culture that had shown activity against staphylococcal and pneumococcal infections. Among the successful trials included in the report was the treatment of a four-month-old girl suffering from a staphylococcal infection, and that of a boy of two-and-a-half and a sixteen-year-old, both with tonsillitis. The Spanish firm that submitted the report on its use—no individuals are named in the submission—recommended either topical or rectal administration.[3] The extraction process involved production of the mould by surface culture, followed by successive phases of extraction and concentration. This pharmaceutical and its extraction procedure were approved for sale in May 1945 for topical use only. It was sold as dry powder in an ampoule with an additional vial of physiological saline solution. The process of manufacture and uses of the product—involving painstaking work with cheap materials—suited Franco's autarchic industrial policy. This economic system of self-sufficiency and limited international trade relies on autonomous domestic production of food, raw materials and manufactured goods, which Franco had championed during the post-Civil War era. Immediately after the Second World War, when calls from the United Nations to impose economic and diplomatic isolation on Spain were heeded throughout the West, this became a virtue out of necessity.[4]

The main reference for the report submitted by the laboratories that produced this penicillin powder was the second paper published by the Oxford group in 1941. It is in this article, which described the setting up of tests and methods to obtain penicillin from the mould by surface culture, where a 'standard ... unit of antibacterial activity' is defined: the amount of penicillin which when dissolved in 1 cm^3 of water and added to a Petri dish containing a *Staphilococcus aureus* culture produces an inhibitory circle (clean circular zone meaning no bacterial growth) of 24 mm diameter.[5] Once this unit was established by the group, it began to be known as the 'Oxford unit' and flasks of penicillin would travel the world with clear information about the doses contained. Through the 1942 paper—a container itself, holding all the suitable recipes and standards—knowledge about the therapeutic properties of penicillin travelled swiftly and easily. The Oxford group—five men and two women—retained credit for their pioneering work; by 1946 two of the men had been awarded Nobel Prizes together with Fleming, who kept his role as initiator of the entire penicillin story. His method for testing a penicillin solution remained in use and his status was maintained largely through strains of the mould Fleming supplied to laboratories across the world.

The artisanal production of penicillin by surface culture was, as stated, fully in tune with Franco's autarchic regime. Political isolation and autarchy made national production a virtue out of necessity. This strategy failed many times, however, as indigenous entrepreneurs made assurances of national production that turned out impossible to fulfil.[6] Autarchy did not mean total isolation, however. Scientific and medical journals appear to have effortlessly crossed the political boundaries that national frontiers were at the time; walls built by Franco's regime to restrict the circulation of knowledge, commodities and people. Passports were difficult to obtain: among the many requirements was that of not having any 'criminal record', which excluded those imprisoned for political and ideological reasons. Authorisation for commodities to cross the border was also difficult to obtain, but knowledge appears to have circulated easily: as discussed in Chap. 1, scientific journal distribution by mail was one major channel.

In 1945, the year penicillin was approved for sale, the drug was the subject of pharmacist Víctor Villanueva's dissertation, presented as he was awarded membership of the National Academy of Pharmacy. An influential expert, Villanueva was head of the pharmaceutical section of the Union of Chemical Industries, and advisor to the Ministry of Industry. In his dissertation he stated that both the chemical composition and fermentation process for penicillin were unknown, such information being kept secret, but that this valuable new drug was produced through fermentation of the mould *P. notatum*.[7] Thus, by this time, clinicians, scientists and the public in Spain were fully aware of the efficacy of penicillin in the treatment of infections. The Spanish view of penicillin was linked to the public imagery of Franco's dictatorship, to the heroic overtones of the heavily censored literature of the time, to the autarchy of his government, and, as was argued in Chap. 2, to the symbolic figure of the hero. Physicians, patients and the public alike perceived penicillin as such a hero. The Spanish authorities began to recognise the necessity of domestic penicillin production, and by 1948, it was officially considered to be of significant interest to Spanish industry. Spanish-made penicillin reached the market in 1952, through contracts with the American firms that first mass-produced the drug.

The people of Spain, the national market and the dictatorship were all agents in the construction of a national medical and pharmaceutical life, all contributing to the inclusion of drug manufacture and medical practices in historical narratives.[8] Rather than being a national icon, as the drug was in France after the Second World War, and in the USA during its mass-production and government-run, large-scale clinical trials, during the early days of Franco's dictatorship penicillin functioned as a heroic medical and scientific object, a state-controlled symbol of recovery, despite being discovered, developed and produced overseas. It shared the political life imposed by the dictatorship while at the same time captivating the public's imagination. For those interested in business the drug was a promising commodity, fully tested in neighbouring national markets. Once ready for the Spanish market, the government incorporated also domestically produced penicillin, as did with the legal batches of foreign penicillin, into state bureaucracy.[9]

The penicillin factory—the space of industrial manufacture—is regarded here as both a physical space and a symbol in the reconstruction of the economic and cultural meaning of the new drug. The factory embodied a system of production, a market and the capitalist economy, with both labour regulations and technology at play.[10] With a manufacturing space for its production, penicillin made the transition from a heroic saviour to a commodity; from a promise to a flask of white power, to be applied following suspension in distilled water using equipment that could be bought in any pharmacy. Its system of production provided the machines and the women and men that handled them. The manufacturing and bottling of penicillin powder in bulk required careful handling, and the following of a detailed set of instructions contained in patents, technical recipes protected by industrial agreements between firms from different countries. Property rights governed a space of technology and workers while gender articulated the distribution of physical and symbolic spaces in the manufacturing line. The bottling plant exhibited this gendered distribution, as did the manufacturing plant itself, as a space and as a process.

Towards National Production

The political negotiations for manufacturing penicillin in Spain involved clinicians, astonished at its effects, and industrial and health authorities, impressed not only by its success but by the demand this success created. Industry was among the most powerful political domains of Franco's cabinet: the Ministry and a National Institute of Industry (INI) owned, promoted and controlled every industrial firm during the 1940s and early 1950s. Immediately following the Civil War, Franco's industrial policy was almost entirely focused on heavy industry, such as transport, power production and communication, largely relating to defence and the military. Health administrators, as part of the Ministry of the Interior—which also governed the police—had less political stature and decision-making power than industrial authorities. With the shifts in relations at the end of the Second World War, Spanish industrial relations with foreign firms—previously isolated both by international diplomatic policy towards the regime and by Franco's autarchy—began to be revived by the banking sector and individual entrepreneurs.

By the mid-1940s, with penicillin receiving adulation from clinicians world-wide, industrial authorities in Spain still had doubts about manufacturing the drug. In the early days of the regime, industrial strategy and political ideology controlled health issues, including the distribution and prospective manufacturing of penicillin. The production of penicillin was demanded by clinicians and citizens, the former through prescription, the latter through a determination to obtain the drug by any means, whether through official channels or illegal trade.

'There is nothing as marvellous as watching a fever decrease and the gravest symptoms of septicaemia, even those caused by a deadly staphylococcal infection, subside a few hours after beginning penicillin injections. Those of us who have seen patients emerge out of insuperable pain can neither complain nor curse our times.'[11] These words by the renowned Spanish clinician, Gregorio Marañón, in the foreword to Florencio Bustinza's first book, published in 1945, echoed Bustinza's own describing both the drug's wondrous effects and the curse of insufficient supply: 'What therapeutic results might be achieved when pure penicillin can be used as a medicine, at an affordable price, at the beginning of an infection and at suitable doses, without restrictions and the fear we have today of wasting it?'[12] Marañón and Bustinza's views represent the Spanish Zeitgeist in regard to penicillin. Its therapeutic activity and scarcity were the bases upon which a legend was constructed. So unusual and unforeseen were its effects considered that the telling of the drug's story took on legendary form. However, disease and pain were not legends and the healing capacity of penicillin and its growing renown determined the agenda: it should be used 'without restrictions' and at 'an affordable price'.

As in other countries during the Second World War, the manufacturing of penicillin outside the USA and UK consisted of small-scale surface culturing of the mould *Penicillium notatum*, by expert microbiologists or amateur scientists, resulting in small amounts of an impure drug. After the war, penicillin manufactured in the USA began to circulate in Europe. A number of firms in France and Germany began production on license, while smaller companies in Austria, the Netherlands, Sweden and Denmark also acquired the necessary methods and technologies.[13] In Poland, Czechoslovakia and Yugoslavia, factories created with initial support from the UN Relief and Rehabilitation Administration had begun to produce the drug by the end of the 1940s.[14]

The increasing demand for penicillin in Spain, along with its scarcity, despite a thriving black market, became so blatant that in September 1946, the Dirección General de Sanidad submitted a proposal for domestic manufacture to the industrial policy authorities. Penicillin consumption was so high, the proposal argued, that the official currency budget for importing penicillin—$5000 per month at that time—was inadequate. Rather than increase this expenditure, the heath authority report considered it 'more logical' to support manufacture of the drug within Spain through an agreement between industrial policy authorities and 'civil industry'.[15] The industrial authorities 'discussed' the subject in January 1947 and the Ministry of Industry received an application to build a penicillin factory in June, together with a memo and report by industrial policy advisers.

This first application for manufacturing penicillin in Spain had been submitted in 1946 by Leopenicilina Española, 'a company from Bilbao'. The firm requested permission from the INI to establish a factory using a patent from the Danish company Løvens Kemiske Fabrik, based in Copenhagen. As Leo was the pharmaceutical division of Løvens, this application appears to have been an attempt by the Danish firm to establish a penicillin plant in Spain. Løvens Kemiske Fabrik had been funded in the early twentieth century by two Danish pharmacists, and initially produced hormones, including insulin. During the Second World War, Løvens' director became aware of penicillin's potential immediately after publication of the original *Lancet* article in 1940 and, in collaboration with local microbiologists, managed to obtain the drug through a surface culture. The firm developed and used their own method of purification until Ernst Chain visited the Danish factory in 1946 and recommended the submerged fermentation techniques developed in Peoria. To achieve this, Løvens built a new factory in Ballerup (Denmark) and developed its strategy for international expansion. This Danish firm went on to manufacture penicillin in Italy.[16]

The accompanying memo, apparently written early in 1947 by the health authorities in order for the INI to evaluate Leopenicilina's proposal, stated that penicillin had achieved wide medical recognition in the treatment of many diseases and its field of application was expected to expand, therefore sufficient quantities of the drug were required.[17] The report expressed doubts about the Danish firm's research capabilities compared to American firms such as Abbot Laboratories, Lilly and Merck. Abbot had already achieved public recognition by publishing the book *Revisión de los conocimientos actuales de la Penicillina* (Review of Current Knowledge about Penicillin).[18] This detailed account of the drug's production, properties and clinical uses, written in comprehensible Spanish, effectively circulated both information about the power of penicillin and awareness of Abbot as a manufacturer among clinicians. *Penicilina* was already being distributed in many South American countries by Abbot, and with chapters such as 'Especialidades de Penicilina Abbott' (Abbott Penicillin Specialties), the book could give the impression Abbott was already marketing the drug in Spain.[19]

The report on Leo's proposal 'hoped' that ongoing improvements in penicillin production might yield both scientific and economic benefits, expressed concern about the high cost of licenses and the flasks penicillin would be sold in by Spanish producers, and suggested that an investigation of the Danish patent and its cost be carried out. A representative from Abbot Laboratories had visited the Dirección General de Sanidad and recommended they obtained more information about Abbot's intentions concerning penicillin production in Spain. According to the report, American imports of penicillin were cheaper than the price proposed by Leopenicilina Española. Although the need to manufacture penicillin in Spain was evident, the report nevertheless recommended building contacts with foreign producers and seeking clearer information about Abbot's intentions in particular. There seemed to be no hurry to establish a factory, or any great interest in domestic production while the need to buy a patent appears to have been taken for granted.

Both Leopenicilina Española's application and the report written about it reveal detailed technical knowledge about methods of production, informal contacts established by pharmaceutical companies with both the health and industrial policy authorities, and about the growing demand for penicillin: at this time medical journals were beginning to report on successful treatments with the drug, alongside news of exciting results obtained with an even newer drug, streptomycin.[20] At least two foreign firms were attempting to enter the Spanish market and begin penicillin manufacture in Spain under their own patents. An expert in penicillin was needed to evaluate the proposal for the health authorities, a small group of industrial politicians and physicians in need of further information. Apparently Florencio Bustinza was performing this role, or trying to, as an intermediary in the preparation of submissions.[21]

With the industrial authorities' focus on heavy industry, production of penicillin was not regarded a priority. In the first decade of Franco's regime, when autarchic self-sufficiency was official policy, the INI was responsible not only for promoting industrial production, but also controlling Spanish initiatives in all industrial sectors.[22] The state was involved in various branches of the chemical industry, including the production of fertilisers, cellulose and artificial fibres, as well as perfumes, and aromatic and resin derivatives; in some instances this assistance was in response to requests from the industrial firms themselves.[23] This period of economic nationalism, while sharing the strategy of promoting national production with many post-war nations, reinforced the dictatorship's isolation.[24] Private initiatives that were able to develop did so under the umbrella of 'industry for national interests', a concept and political practice regulated through a decree that came into force at the end of the Civil War; many private initiatives were explicitly discouraged.[25] It was from within this heavily controlled landscape of industrial and heath care practices that Franco's government finally decided to promote the establishing of two domestic penicillin manufacturing plants.

Penicillin Manufacturing as a National Interest

Three months after Fleming's visit to Spain in June 1948, and his reception of one of the highest honours the Spanish government could bestow, the *Gran Cruz de Alfonso XX el Sabio*, the industrial authorities finally responded to public demand for penicillin. The decision to allow the creation of penicillin factories in Spain, granted to two consortia in August 1949, followed lengthy discussions and negotiations involving clinicians and heath authorities. This shift in attitude owed much to the public displays of adoration directed toward Fleming and, therefore, toward penicillin.

In September 1948, the Ministry of Industry approved a decree that declared 'Penicillin: its manufacture is of national interest'. When this was published the following month, an open public competition for the right to establish two factories in Spain was launched.[26] By considering penicillin production to be of national interest, the authorities assumed control over its manufacture and finally included penicillin in the government's industrial priorities. The decree required firms to provide information about patents and technical advice, which it was assumed would originate from foreign companies: the proposals were asked to specify any need for foreign currency. They were also required to give details about the origins of the technical material and devices requested, as well as a detailed market study. Economic relations with other countries were deemed necessary to benefit from the therapeutic activity of penicillin. Although closing the borders to foreign products was the official policy of Franco's regime during the 1940s and early 1950s, this coexisted with the importation of products and tools for industrial manufacturing. Total isolation was impossible to maintain, despite the ever-present political rhetoric of autarchy. Regulations did remain strict, however, resulting in long delays in the release of both foreign goods from customs and the foreign currency required for payments.

It was shortly after Franco's conquering army entered Madrid in October 1939, an event that marked the end of the Civil War, that the government had published the decree on the provision of support to industries 'relevant for national defence and the economy'; the first major regulation of Franco's autarchy. This support included a tax reduction of up to fifty per cent and reduced customs duties.[27] In 1948, it was announced the government would extend this support to the firms selected for penicillin production. Spain was a virgin market for penicillin at a time when a new antibiotic, streptomycin, was already being manufactured. A number of industrial groups and pharmaceutical laboratories were interested in what appeared a promising business.

By that time penicillin was regarded by the recently created World Health Organisation (WHO), the health branch of the United Nations, as 'the new anti-syphilitic drug'. In April 1948, immediately after the WHO constitution came into force, its Committee on Venereal Diseases—one of the earliest to be established—reported on the treatment of syphilis and other venereal diseases with penicillin. The report encouraged the production of penicillin to ensure an 'equitable distribution' of the antibiotic to all countries and also included, 'on the basis of available data', a recommended dosage to be taken every two hours over eight days.[28] This suggests a constellation of circumstances played a part in the Spanish government's decision to promote domestic penicillin production. It also reveals the agency of the drug's ability to cure venereal diseases in the history of penicillin.[29] The constellation included the public silence regarding syphilis and venereal diseases in general—both in archival material and the regime-controlled press—and the government's desire to be accepted by the new international agencies being created by the Allies. Such a landscape illustrates the interplay between domestic politics and international affairs that prompted the reconstruction of public health policy in Spain, an early measure of which was the decision in September 1948 to authorise the two penicillin manufacturing plants.

With the aim of establishing contacts with Merck & Co. to prepare a proposal for the competition, the young professor of physiology at the University of Madrid Medical School, Antonio Gallego, travelled to Rahway, New Jersey in January 1949. He represented an industrial chemical and pharmaceutical group, of which his brother José Luis Gallego was a director, belonging to the Spanish bank, Banco Urquijo. Antonio Gallego had spent the war in clinical service at the Republican front in Madrid. When the war ended, under advice from his brother, he voluntarily entered a concentration camp in El Pardo (Madrid) to avoid retaliation; through family connections with the authorities he was released some months later. Antonio Gallego's 1944 PhD dissertation had been on the structure and function of the retina, which became his research topic as a physiologist. As a postdoctoral fellow he then went to the Rockefeller Institute of Medical Research in New York to work with the neurophysiologist, Rafael Lorente de Nó, before returning to the University of Madrid in 1947. His stay in New York and close family ties to Banco Urquijo probably explain why he was assigned to represent the firm in negotiations with Merck.[30] In January 1949, Antonio Gallego signed an agreement to establish a penicillin factory in Spain with Merck's vice-president, R. E. Gruber.[31]

The agreement included a commitment from Merck to assist, advise and provide detailed information about the fermentation process and technical requirements of a penicillin production plant, 'if BU [Banco Urquijo] is granted such a concession'. Merck would also supply suitable cultures of microorganisms, as well as initial nutrients and a full description of the process.[32] Technical details, including all information necessary to prepare the bid for the Spanish government, were regarded as 'strictly confidential', and not to be used for any other purpose; this secrecy was absolute, and not a single document has been located containing this information. The agreement letter mentions the use of a 'deep fermentation method' to produce 'sterile crystalline sodium Penicillin G [benzylpenicillin], sterile crystalline potassium Penicillin G, sterile crystalline procaine Penicillin and other kinds of penicillin regularly manufactured on the date BU is granted such concession by the Spanish government'. Between March and June that year, Merck submitted four patent applications: all were approved by the Registro de la Propiedad Industrial (Spanish Industrial Property Rights Registry). At the end of 1949, the US firm Schenley Industries, Inc. also submitted four applications. As Ana Romero has insightfully shown, approved patents were published in the *Boletín Oficial de la Propiedad Industrial*, but also had to appear in a widely distributed daily newspaper and a trade newspaper or industrial journal. Patenting procedures therefore included making technical production procedures public knowledge while keeping such procedures under legal protection.[33] Thus, patenting was the end of secret knowledge.

In January 1949, the Academy of Pharmacy claimed that its members, as experts, should be consulted in the evaluation of any factory proposals. Members of the Academy were cooperating with or working for pharmaceutical firms, including a number that had submitted applications, and pointed out that with the 'reserved' technical details omitted, it was extremely difficult to properly evaluate the proposals.[34] Bustinza, an Academy member, did participate in an evaluation report, apparently at the authorities' request. He had contacts with Leopenicilina, but the proposal submitted by the Danish firm was not approved.[35] The Academy's demand shows the interest the open call had inspired, and how the system of professions intervened, or attempted to intervene, in a process in which medical and pharmaceutical interests were at play. As scientists, as expert pharmacists, they attempted to negotiate with the industrial and public health authorities, to establish their own role in the approval process, and gain recognition of their scientific authority.

Two Antibiotic Factories

In August 1949, the government approved two proposals. One was a joint submission from Consorcio Químico Español SA, a group of chemical and pharmaceutical companies, and Banco Urquijo, who were authorised to establish a penicillin factory in Madrid. Approval to build a factory in León was awarded to Industria Española de Antibióticos SA (from now on, Antibióticos SA), a group of six small Spanish pharmaceutical firms—Abelló, IBYS, Zeltia, LETI-UQUIFA and Instituto Llorente—with experience in the production of sera and vaccines before the war. Both had to guarantee Spain's projected requirement of penicillin for fifteen years, and a minimum capacity of six million doses of one hundred thousand Oxford units per year. Any contracts signed with foreign firms had to be submitted to the industrial authorities for approval, and material imported for the manufacturing process would be controlled by the state, at a time when imports were severely restricted due to the lack of foreign currency.[36] Both groups were private firms entering into agreements with foreign companies—Consorcio Químico Español SA with Merck, Antibióticos SA with Schenley—reinforcing the fact that autarchy was never completely established. Although economic and industrial protectionism had been practised in Spain and many other countries in previous times, during the first decade of the dictatorship, autarchy was yet another iconic representation of Franco's government. From the late 1940s on, however, the purchasing of foreign patents was a customary feature of Spanish industrial life.[37]

Although not a requirement for proposals to be submitted, Antibióticos SA requested state participation in the venture. Assessing the full commitment and manufacturing capacity of Schenley, Antibióticos SA's share proposal insisted that Schenley was among the best penicillin producers 'in the world', and one of the first involved in production. Schenley had recently installed a factory to produce 'Penicillin G' in Paris for Sofrapen (Societé Française de Pénicilline), which only a few months after opening was producing a greater quantity of penicillin than projected. INI, with its duty to promote large national industries, was considered to be the 'best partner' by Antibióticos SA: the consortium may have been looking for the guarantee of 'absolute preference' in the use of new methods.[38] In November 1949, however, a technical report was published which regarded this partnership unsuitable. INI, the report argued, could only enter into partnership with a firm that lacked financial capacity, which was not the case for either firm.[39] This was not a hard and fast rule, however, as the state was involved in other industrial domains with firms that needed no financial backing, and contributes to the suggestion made above about the low consideration the industrial political authorities gave to the antibiotics business and its social impact in public health and medical care.

Meanwhile, in October 1949, after a decree was passed to authorise their project, Banco Urquijo created a new firm: Compañía Española de Penicilinas y Antibióticos (CEPA). Thus two companies, CEPA and Antibióticos SA, would share penicillin production within Spain—and later the production of other antibiotics—with patents from US firms.

Gendering the Manufacturing Plant

As part of their contracts, both companies were required to establish plants for bottling imported penicillin while their production plants were under construction. CEPA took seven months to build their bottling plant in Madrid, starting production in August 1950 (see Fig. 3.1).[40] Merck's Vice-president, J. Korrigan, travelled from Rahway for the inauguration; his eulogising comments on CEPA's technical facilities were published in newspapers as a two-page advertisement.[41]

Other advertisements for penicillin had begun to appear in the Spanish press, the vast majority featuring photographs of women seated at machines, weighing, bottling and labelling penicillin. Women workers were also included in news reports on this modern national industry, suggesting the theme was deliberate, to emphasise women's participation as industrial workers to promote Spain's antibiotic industry.

'Spain is going to have enough penicillin to cover its needs', proclaimed a double page of publicity in September 1950.[42] In the summer of that year, as CEPA and Antibióticos SA started bottling penicillin imported in bulk from the USA, the firms also began their advertising campaigns, rallying interest in their industrial endeavour to manufacture the wonder drug of the day. The campaigns included advertorials, with extensive text and two large photographs of women at work. The pictures in CEPA advertisements are extremely similar to those included in news reports, suggesting the firm had carried out promotional work to boost coverage. These images are the sources I have used to reconstruct how the bottling plant and its technology were represented and gendered.

These photographs were presented to the public at the end of the miserable, poverty-stricken decade following the Civil War as a visual rhetoric of national achievement. Antibiotic production is depicted through women sitting at bottling benches and their participation in innovative technical practices. The women wore white coats and caps covering their hair like nurses; factory nurses taking care of the new

Fig. 3.1 CEPA bottling plant in Madrid. Courtesy of Roberto Gallego

wonder product that required safe conditions for its bottling. They wear uniforms; they are standardised and thus, as standards, remain a collective of dissolved individualities, as in many production-line assemblies. It is from this collective that they provide an identity based on their role as industrial workers.[43] These women are not looking at the camera but are in a working stance, heads bent forward, carefully examining the contents of the glass cases: according to the wording alongside, this was pure, sterile penicillin powder. All are bent at the same angle, a uniformity that exhibits safety and order. A perfect similarity between the women in the photograph makes them not only equal but also anonymous, indistinguishable from one another.

A woman's hands were also shown. Two hands wearing gloves 'as if surgeons', demonstrate the central role of the precision scale, one of the technical devices also praised in the advertisements, and the leading role of which reinforces the precision of the women's work. According to the text, a sterile environment was of major importance in the penicillin purification and bottling process: women took compulsory showers and adhered to other strict hygienic measures, suggesting the purified environment these women should provide.[44]

'To be really efficient, penicillin needs two conditions: absolute purity and exact quantities', appears in the account of women's tasks in the factory.[45] Praising women was praising cleanliness and safe handling; that is, praising penicillin itself. The representation selected to advertise the drug was composed of women dressed in white, in front of stainless steel boxes and purified flasks. The impact was doubly symbolic: of the drug and its wondrous effects, and of the women's work while handling it safely in this innovative, technical environment. The women workers were segregated, emphasising their selection for gendered work in an apparently stable social order.[46]

These images exhibit women attached to the new, innovative devices designed for bottling and sealing the drug. The set—a women worker at a machine—is presented as an innovation in safety for manufacturing a product to therapeutically provide cleanliness and purity. The women's images represent this promotion of the new drug: their uniforms were washed and ironed daily, and the environment kept clean by automatically closing double doors. Workshops were located in small rooms for the same objective, and hermetically sealed, stainless steel cases were provided, with two germ-killing ultraviolet lights and filters for the air circulating inside. Gloves were used to handle the products inside the glass cases.[47]

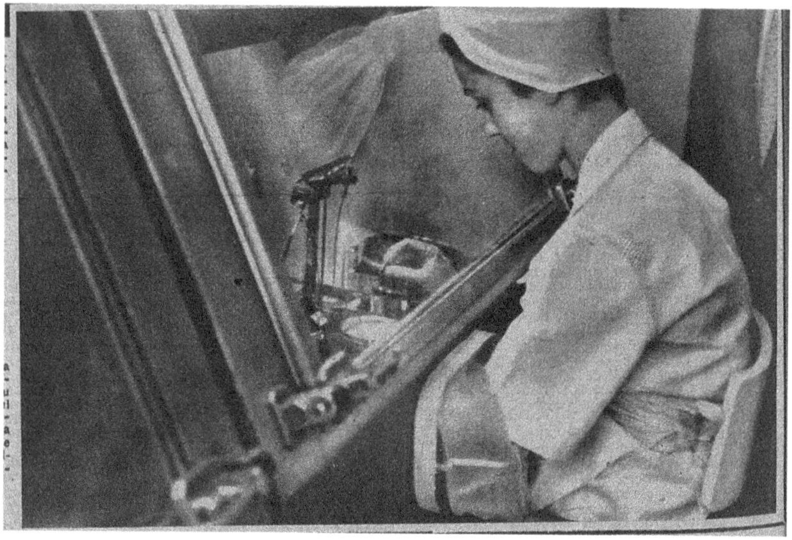

Fig. 3.2 Woman weighing penicillin at CEPA bottling plant, Madrid, ca. 1950. *Source ABC* Madrid, August 16, 1950, 4

These photographs (Figs. 3.1 and 3.2) are similar to those of women factory workers taken in other countries, in the manufacturing processes of medicines before penicillin as well as the production of many other products and services. This was despite the post-Civil War politics and norms of the dictatorship having sent women back to domesticity. Although this was happening in other countries after the Second World War, as men returned to their professional spaces, some of which had been occupied by women during the war, sex segregation under the dictatorship was particularly distinct, with women losing some of the rights they had achieved and kept legally dependent on men: fathers, brothers or husbands.

While a small number of women did work outside the home it was the low salaries men received during the post-Civil War years that led many more women to search for work despite the gender policy of the dictatorship. In addition, many women-centred workplaces were created in the post-war days, as had happened before, such as cigar production at the turn of the century. Older than the scientific revolution and the successive industrial revolutions, women cigar makers remained in their workplaces all through the technological transition from handmade cigars to mechanically produced cigarettes.[48] Historian Lina Gálvez argues that it was the flexibility exhibited by women that kept them working in the cigar factory and tobacco production in general. Technological innovation was not fully mechanised, and 'the development of hand and intermediate techniques', for which women had displayed particular talent, were still required. The same flexibility has been observed in the early decades of the Spanish telephone company, Telefónica, as new workplaces were created for women telephonists.[49] These jobs were created by telephone companies all over the world, and women operators have been active protagonists in the technological changes within telephonic equipment, from the early days of manual connection to automated communications.

Also, and more relevantly for women factory workers, the processes of industrialisation include, as Maxine Berg has phrased it, 'the wider use and division of cheap labour'.[50] Women, and in earlier times children, have often functioned as cheap labour, with some technological innovations designed specifically for this vast resource. As women's training was rarely carried out in technical institutions, it was often regarded as 'informal', and thus cheap.[51] The workplaces they have inhabited are deemed not to require particular training, only the flexibility already mentioned. It is to this history of women workers that the occupational identity of the penicillin bottling plant workers belongs.

This gendering of particular types of industrial work, such as attributing the safe handling of sensitive, delicate products on a production line to the traits of women, was promoted in the Spanish penicillin advertising campaigns. At the end of the first miserable decade of Franco's rule, a high number of young, single women were able to fit both the requirements defined by the (male) managers of a factory line and the political demands that kept married women in the home. Many women became married and had children while keeping their jobs, despite a social order reinforced by the laws of the Franco dictatorship and its representation of masculinity in the form of the army's hierarchical order: Franco as generalissimo of the Spanish army, navy and air force, who governed the country through its recovery from the miseries of the civil war and the post-war years of the 1940s, up until 1975.[52]

The extent of poverty in Western societies, together with trends toward industrial recovery and development, brought women to work in the factories. The contemporary construction of what it meant to be a woman was appropriate for factory work: conscientious women workers handled pure products and used small, new factory devices with care. Women were included among the factory workers in the two penicillin production plants from the late 1940s on, taking part as conscientious technicians.[53] They were thus regarded as reliable contributors to the manufacturing line, in the handling of new, delicate technology, and, in so doing, became part of the active population available for hire by a growing national industry.[54]

Between 1940 and 1950, a decade in which fertility rates dramatically decreased, forty-five per cent of all Spanish women were between the ages of twenty and forty. As the number of single women decreased, the number of women in the workplace increased. This suggests that single women, once entering the labour market, were highly likely to remain there after marriage and childbirth.[55] Women entered the labour market while young and single, when political norms dictated married women should be fully dedicated to their work at home. Among these single women were the 'working girls' at the bottling tables in penicillin production plants, which by the end of 1950 had processed more than a million doses of the drug.[56]

The photographs presented here attest to the origins and trajectories of a social life segregated by sex, which naturalised and normalised differences fuelled by industrial life experience. This sexual division is part of contemporary culture and its narrative, thus illustrating the strength of a system of symbols and cultural values bisected by a sexual classification that became a cultural dichotomy. Images of many earlier penicillin manufacturing plants, such as those published by Abbott in the firm's Spanish edition of 1945, in the archives of Schering AG and in Podolsky's reconstruction of the history of drug resistance, show young women at work without any accompanying comment regarding the gendering of industrial workplaces. In all these pictures women are seated within the bottling, sealing and packaging section of manufacturing plants,[57] and quality control laboratories,[58] in the otherwise male industrial space of the factory.[59]

The audience these images addressed was composed of people who should feel safe consuming a previously scarce, healing product. Following a decade of poverty and a lack of medications, photographs of the penicillin factory depicted modernisation and prosperity. Here was a direct promotion of penicillin, assuring the public the drug would be safe and no longer in short supply: 'a major problem has been solved' declares the advertisement.[60] By the end of 1950, CEPA publicity could boast of producing the first half million flasks of Penicillin G (see Fig. 3.3). During 1951, CEPA bottled 12 million doses of 100,000 units, the maximum amount projected by the decree.[61]

The Factory and Social Order

CEPA opened the construction of its production plant in the town of Aranjuez near Madrid. A decision was reached in March 1950, building work started in July and the installation of technical machinery began by December. In September 1951, CEPA started to produce penicillin, exactly—according to its 1952 annual report—two years after being granted manufacturing rights. The technical director of CEPA was a pharmacy graduate, Carmela Viejo, a women participating within the hierarchical management of the plant. The image of Carmela Viejo that has been preserved shows her white-coated, guiding factory managers through the workshops and aisles (Fig. 3.4). She was regarded as a highly competent director, skilled and smart, who achieved everything she set her mind to.[62] We see her with the directors of the company, Justo Martínez Mata and Antonio Gallego, visiting the basements of the bottling plant, surrounded by the plumbing of the sterile facilities. Although she was part of the same job category as Martínez Mata, she is wearing a white coat, as technicians did, the term used for university graduate workers in the factory research laboratory. In her case, white conveyed educational level and scientific recognition. Although unusual at that time, a woman as technical director did reflect the high number of Spanish women who had studied pharmacy since the early twentieth century.[63] Carmela Viejo and her position at the factory illustrate the extensive presence of women at many levels within social and professional domains; women who have not been studied or attributed leading roles in the historiography and thus join the invisibility of women workers. Carmela Viejo has become visible through photographs preserved by the family of CEPA director, Antonio Gallego.

Men are also depicted in these images of the penicillin factory, wearing ties and on foot. Their male clothing and standing position at the factory door encodes the power of a male hierarchy that selected the representation of men as authority—Martínez Mata, Robert and Gallego had all taken part in the complex negotiations to create the factories—and women working in the bottling line as user participants in technical

Fig. 3.3 CEPA advertisement 'Medio millon de frascos'. *Source ABC*, January 31, 1950

Fig. 3.4 Carmela Viejo, white-coated, with Antonio Gallego and one of the directors of CEPA, on the right. Behind, Justo Martínez Mata. Undated. Courtesy of Roberto Gallego

innovation. Male workers were portrayed in factory photographs wearing blue overalls standing at the foot of fermentation tanks, handling the stopcocks and main taps of the circulating liquids and gases.[64] Men were in charge of the section containing large machines: steam boilers, air filters, devices for water depyrogenation (removal of pyrogens or substances that can produce fever, mostly bacteria) and deionisation, steam pressure reducers, and detergent tanks. 'Each device is a pattern of perfection and modernity', proclaimed the news report.[65] The size and shape of the new buildings were also emphasised, as part of the expanding industry antibiotics manufacturing was becoming: property as power.

This gendered distribution of tasks, which was also a distribution of space, evokes older associations of women with caring tasks and duties that involved the handling of small devices: caring for children, the sick and the aged, as well as providing food and daily remedies beyond their families and households. Taking a feminist approach, images of women in white and men in blue in the factory can be seen as publicising a genderised industrial development.[66]

By March 1952, CEPA was producing a higher output than required by the decree. In June, monthly production reached one million doses of 100,000 units. This amount was still 'insufficient to face market needs' as the bottling plant had been producing double the required number of doses per month during the whole of 1952, according to CEPA's annual report. This situation had already been foreseen, and for this reason, CEPA submitted a request to the authorities in November 1950 for a hundred per cent increase in penicillin production.[67] By 1953, CEPA was producing 24 million doses of 100,000 units per year, on a working regime of 48 hours per week, and was proposing doubling or even tripling this production rate by instituting supplementary work shifts and using reservoirs to increase fermentation capacity. Once established, this regime enabled CEPA to cease importing penicillin: production by themselves and Antibióticos SA was able to cover national demand for the drug. By that time, a CEPA streptomycin plant was under way, working at a 'satisfactory' rate.[68]

Gendered Penicillin and the Politics of Industrial Production

By acknowledging the segregated contribution of women and men in this history of penicillin manufacturing in Spain, the relationship of women and men in industrial history becomes visible, as do the connections between past and present social practices regarding differentiation of the two sexes.[69] As part of the industrial West during the post-Second World War era, the Spanish government was laying foundations for the economic expansion of the 1960s. In the penicillin factory, at the bottling tables of sterilised workshops, gender circulated; individual women and men shouldered this category. Factory images are taken here to be representations of the industrial setting in which gender condensed as a category and unveiled itself as dynamic. Gender invaded the space of practices, and women have been made visible, both by the industrial environment of the time and through my own narrative.

Experiences as collective practices have been structured within political and geographical settings, where time and place have played a part. This gender circulation broke the norms regarding women's place in society; norms established by the Western biopolitics of the time, which extolled families, maternal love and women as articulators of affection in the fundamental social structure the family had become for post-war recovery. Women, however, became an increasing part of *human resources* for working outside the household. Their participation in the antibiotics factory and the research laboratory of the factory, as conscientious workers privileged by gender categorisation, illustrates their agency in producing changes in the factory, and society at large. Women participated in the industrial space being created.

These images reveal the impact of women's presence on antibiotic production, as a recognised collective in white that represented prosperity. By recognised I mean they were represented in the images because women or, more likely, their workplaces in the bottling plant, were worth exhibiting. Their worthiness meant these images became agents for the credibility of the factory, for what was manufactured there and for the tasks involved. This was part of the wonder of penicillin: the values attached to it in circulation. Attached to penicillin, gender as a category was also moved, mobilised and carried by penicillin propaganda.

Men have been regarded and studied as the main characters—directors, authorities, negotiators, politicians, engineers, clinicians and professors—in the history of antibiotics. These images in an antibiotic factory represent women's social condition, read by juxtaposing text and pictures and their historical and geopolitical location. Biological models—sexualised bodies—became embedded in lifestyle and social roles, and were agents in the construction of gender as a cultural category capable of being analysed.[70]

The hierarchies of the work place, transferred to the historical reconstruction of industrial production, are represented by the distribution of physical space at the factory and the different symbolic and physical spaces segregating the sexes. This separation in penicillin manufacturing took the form of the social and cultural relationships presented here: men wearing ties in the dark suits of white collar workers or in blue overalls standing beside large fermentation tanks; women dressed in white coats and caps, like sanitised nurses in the purified atmosphere and small-scale technology of the bottling workshop. Through examination of the spaces they occupied within the factory, a social order emerges: collectives of women and men, and the distribution of recognition, agency and authority between them.

Men were portrayed as directors, in the leading roles of either economy or politics; women were selected as strategic representations of prosperity and the modernisation of early antibiotic production. By creating these gendered workplaces in the factory, CEPA and Antibióticos SA followed previous models from US penicillin factories and German drug factories in general. Women were given roles in the finishing processes, providing credibility and demonstrating the innovative techniques of the factory's facilities while, as a group, remaining a set of shared yet dissolved identities.

Images of women at benches, operating manual devices and monitoring automatic processes, brought attention to innovation and modern production: gendered values were shown to be associated with penicillin, as if penicillin condensed women's symbolic values, as if it was also female. To some extent, gender circulated toward the meaning of penicillin itself. It was a caregiver, a saviour, a clean, pure final product from the manufacturing line able to cure; a delicate task associated with the way women provide care. Therefore the rhetoric of a woman's place being the home was not entirely contradictory with women's work in the factory. As reliable agents of purity and cleanliness, the latitude afforded these working women only reinforced the political discourse of women's limited rights, by letting them leave home to do what they, according to the norms of the time, really knew how to do: provide care. With their aptitude for purity and cleanliness, women fitted into the social order while also breaking it.

By questioning the construction of women's identities and the systems of representation that have defined and stabilised the gendered structure of antibiotic production, within the contemporary system of the production of goods and knowledge and the culture at large, a space in penicillin history has been retrieved here in which the dynamics of gender circulated. In the late 1950s, along with governments across Europe, the dictatorship was attempting to increase national production. Changes were introduced as women's presence in the workplace increased, but male authority remained in both visual representations and managerial practices, even if, as was the case, women's individual and collective identities as workers and social agents were permanently reconstructing order in a sex-segregated space of activity.

Industrial Production of a Gendered Drug

The adversity faced by society during the early dictatorship created a space of intervention in which the authorities, faced with an ever-increasing demand for penicillin, permitted and assisted private enterprises. The industrial and public health authorities came together to address the problem, at a time when the former had far greater power. After the Second World War and six poverty and terror-stricken years of the dictatorship, there was still space in the regime's political economy to give two pharmaceutical consortiums the opportunity to conduct business; to produce a miracle drug in the midst of misery. It is unlikely any of those involved in the entrepreneurial adventure of establishing a new sector within pharmaceutical manufacturing were fully committed to any social aim: the information available suggests that even if firms obtained some political and financial support, the manufacturing of the drug—the effects of which had been acclaimed throughout Spain during Fleming's visit the previous year—was not a priority in the regime's industrial policy. However, penicillin did become available, as would domestically produced streptomycin in the mid-1950s. The 1948 WHO report and international appeal to face the challenges posed by venereal diseases and syphilis in particular, would have had political influence, despite the regime's particular position toward these infections—both secret and blameworthy—apparent through the silence regarding venereal diseases in public documents and archival material.

Selman Waksman, the discoverer of streptomycin whose research group at Rutgers University was being funded by Merck, visited Madrid for the opening ceremony of CEPA's factory. Waksman's trip was not as celebrated as Fleming's: there was no comparable media coverage or display of citizens' gratitude, suggesting the period of both scarcity and heroic representation was largely over. The passing of both leads me to believe they were linked: the heroic reception of both penicillin and Fleming was a response to the misery and political repression of the early dictatorship. The relief penicillin provided was not only from illness; a business opportunity was generated that produced benefits from its inception, carrying the promise of a better life in its fame and its materiality.[71]

The first factories for penicillin production were built in Spain when public health, US business leaders and Spanish pharmaceutical and chemical firms joined forces. Clinicians, university professors, political authorities in industry and public health, the banks and economic authorities that permitted payments in foreign currency, all were the basis of a new space of business that connected these domains. Penicillin was not only a medical drug and an industrial commodity but a material around which negotiations between different political domains took place. In a situation of limited civil rights and no transparent procedures, and in the midst of urgent demand for penicillin, the landscape was even more complex. Additional agents in the early history of penicillin include the driving forces of advertising and marketing.[72]

When advertising is brought into focus, women appear. The workers portrayed in the advertising campaign announcing Spanish production of penicillin display a social order in which gender articulated hierarchy in the workplace. Historical reconstructions of major events generally situate the pioneers, the directors, the inventors—mostly men—at the forefront of their narratives. Indeed, every name that appears on the proposals submitted for establishing the penicillin factories and the documents awarding authorisation is that of a man.

In the introduction to his book *Fear and Progress*, historian Antonio Cazorla describes a poor family in the immediate post-war years in southern Spain. The mother, providing dessert for her family, is required to distribute two oranges between nine people. The father receives three segments, the son-in-law two, and the rest of the family—the mother, four girls and three boys—share one orange between them. As Cazorla states, 'the harshness of life for people seated at this table was the product not just of their poverty but of politics as well'. Poverty and politics were deeply intertwined during the 1940s, a period Cazorla qualifies as 'extraordinary' hard times for 'ordinary Spaniards'. Women and men had to work outside the household, and factories were one of the places where jobs were available.[73]

By using advertising photographs as sources a collective history can be written, which takes into account agents that are often overlooked: factory workers and particularly those often invisible women workers and their gendered workplaces. Women in white coats and men in blue overalls can be read as a history of antibiotics that segregated the sexes from the beginning, and the promotional strategies of which chose women to represent technical innovation and the cleanliness of an industrial process that offered a new—purifying—medicine: penicillin.

In the finishing stages of production—bottling, sealing and labelling—women were gentle representations of the clean, exquisite manners of modern innovations in penicillin manufacturing, interchanging skills with the drug: the caring, female penicillin.

Notes

1. According to Howard Florey, penicillin was recycled from the urine of patients already treated with the drug in Oxford in 1941: see Hobby, *Meeting the Challenge*, 67.
2. E. Ortiz de Landázuri and J. Perianes, 'Normas generales de la terapéutica penicilínica', *Revista Clínica Española* 17(5) (1945): 361–72.
3. 'Penicilina-OM Tópica. Características y referencia clínica', in 'Memoria para el registro de la Penicilina-OM Tópica, Sociedad General Farmacéutica, Espulgas de Llobregrat, "Extracto de penicillium notatum conteniendo 1.000 unidades Oxford de penicilina por ampolla."' Archivo General de la Administración (Alcalá de Henares, Madrid; hereafter AGA), Ministerio de Gobernación, DG de Sanidad, Inspección General de Farmacia, Sección de registros Farmacéuticos, (8)8 44/18155, 6479.
4. There is an immense historiography on Spanish autarchy, and debate on the extent to which the policy ultimately contributed to or hindered economic development in particular. For an accomplished and respected example, see the approach by historian of agriculture, Carlos Barciela, in the introduction to his edited volume, *Autarquía y mercado negro: el fracaso económico del primer franquismo* (Barcelona: Crítica, 2003). More on this subject in Chap. 4.
5. E. P. Abraham, E. Chain, C. M. Fletcher, A. D. Gardner, N. G. Heatley, M. A. Jennings and H. W. Florey, 'Further observations on penicillin', *The Lancet* 238 (1941): 177–89. The earlier one was E. Chain, H. W. Florey, A. D. Gardner, N. G. Heatley, M. A. Jennings, J. Orr-Ewing and A. G. Sanders, 'Penicillin as a chemotherapeutic agent', *The Lancet* 236 (1940): 226–8.
6. Santiago López García, 'El Patronato Juan de la Cierva (1939–1960). Parte II. La investigación científica y tecnológica', *Arbor* 162 (2013): 1–32.
7. V. Villanueva Vadillo, *Factores determinantes de un proyecto de industria químico farmacéutica nacional, discurso leído en la toma de posesión como académico de número* (Madrid: Real Academia de Farmacia, 1945).
8. On the role of penicillin in constructing national systems of research, see Jean-Paul Gaudillière and Bernd Gausmeier, 'Moulding national research systems: The introduction of penicillin to Germany and France', *Osiris* 20 (2005): 180–202.
9. On the French and British cases, see also Viviane Quirke, 'Penicillin and post-war reconstruction in Britain and France: Science and industry in the balance', *La letter de la Maison Française d'Oxford* 6 (1997): 87–98; Jonathan Liebenau, 'The British success with Penicillin', *Social Studies of Science* 17 (1987): 69–86; and Jean-Paul Gaudillière, *Inventer la biomédicine* (Paris: La Decouvert, 1987), Chap. 1.

10. Inspired by Pat Hudson, Maxine Berg and Michael Sonenscher, eds. *Manufacture in Town and Country Before the Factory* (Cambridge: Cambridge University Press, 1983), Introduction, 3–4.
11. 'Acaso ninguno de esos prodigios pueda compararse con el de ver descender la fiebre y remitir los más graves síntomas de una septicemia, como la estafilocócica, que era inexorablemente mortal, a las pocas horas de empezar las inyecciones de la Penicilina. Los que hemos visto surgir todo esto del caos del dolor invencible, no nos podemos quejar ni maldecir nuestro tiempos'. Gregorio Marañón 1945, 'Prólogo', in Florencio Bustinza, *De Pasteur a Fleming* (Madrid: Editorial Plus-Ultra, 1945) (my translation).
12. '¿Qué resultados terapéuticos se lograrán cuando se utilice como medicamento la penicilina pura, a un precio asequible a todas las fortunas, al comienzo de las infecciones y empleándola a dosis convenientes, pero sin restricciones y sin temor a la responsabilidad que hoy indudablemente tenemos de malgastarla?' Bustinza, *De Pasteur a Fleming* (my translation).
13. Jean-Paul Gaudillière and Bernd Gausemeier, 'Molding national research systems: The introduction of penicillin to Germany and France', *Osiris* 20 (2005): 180–202; Daniele Cozzoli, 'Penicillin and the European response to post-war American hegemony: The case of Leo-penicillin', *History and Technology* 30(1–2) (2014): 83–103, https://doi.org/10.1080/07341512.2014.902232.
14. Sławomir Łotysz, 'Controlling the production and distribution of drugs in communist Poland', *Medicina nei Secoli: Arte e Scienza* 26 (2014): 519–56.
15. 'Moción de 11 de septiembre de 1946', Ministerio de la Gobernación, Dirección General de Sanidad, copia con registro de entrada en el Instituto Nacional de Industria de 8 de enero de 1947, Archivo Histórico del Instituto Nacional de Industria (Madrid), legajo 270, documento 7 (1946–1947).
16. For the history of Løvens, see Cozzoli 'Penicillin and the European'; M. Capocci, 'A chain is gonna come: Building a penicillin production plant in post-war Italy', *Dynamis* 31 (2011): 343–62.
17. 'De Sección del Plan a Dirección Técnica. Asunto: Petición de la Sociedad Leopenicilina Española SA para implantar una fábrica de penicilina', Madrid, 1 de julio de 1947, Archivo Registro General del Instituto Nacional de Industria, División de Industria, Industrias Varias, 'Leopenicilina Española SA', Legajo 200 19/06/47–21/7/47. Archivo del Instituto Nacional de Industria, Madrid.
18. Anonymous, *Revisión de los conocimientos actuales sobre la Penicilina* (Chicago: Abbot International Laboratories Co., 1946).
19. See Anonymous, *Revisión.* Unpaginated but between 152 and 153, inserted as if advertisements. In the first pages it is stated that *Fungicin*,

which was an 'extracto partialmente purificado' was 'in Bolivia, Colombia, Costa Rica, República Dominicana, Ecuador, Panamá, Paraguay, El Salvador, Guatemana, Honduras, Nicaragua, Perú, Venezuela'.

20. Federico Tello Valdivieso, 'Aportaciones al conocimiento de las micosis bronco-pulmonares', *Revista Clínica Española* 20 (1946): 1–24.
21. Cozzoli, 'Penicillin and the Europeans'.
22. Formulario de la gerencia a Presidencia, Petición de la Sociedad 'Leopenicilina Española SA' para instalar una fábrica de penicilina, 21.7.1947, Archivo del Instituto Nacional de Industria, Madrid, División de Industria, industrias varias, legajo 200, documento 7 (1946–1947).
23. Pablo Martín Aceña y Francisco Comín, *INI. 50 años de industrialización en España* (Madrid: Espasa Calpe, 1991), 251–9.
24. J. L. García Delgado, 'La industrialización y el desarrollo económico de España durante el franquismo', in Jordi Nadal, Albert Carreras y Carles Sudriá, eds. *La economía española en el siglo XX. Una perspectiva histórica* (Barcelona: Ariel, 1987), 164–89; Fernando Guirao, *Spain and the Reconstruction of Western Europe 1945–57* (London: Macmillan, 1998).
25. This was precisely the case for early automobile production in Spain; see Elena San Román, *Ejército e industria: el nacimiento del INI* (Barcelona: Crítica, 1999). On 'Industry of national interest', see Ley de 24 de octubre de 1939.
26. *Boletín Oficial del Estado*, October 6, 1948.
27. *Boletín Oficial del Estado*, October 25, 1939, 'Ley 24 de octubre de 1939 (Jefatura del Estado). Protección y fomento de la industria nacional a industrias que se declaren de interés nacional'.
28. 'Expert Committee on Veneral Diseases: Report on the First Session', *Reports of Expert Committees to the Interm Commision*, Geneva, April 1948: 60–end, on 63–4. I am grateful to Esteban Rodriguez Ocaña who provided me with this material from WHO archives.
29. This is an issue that Harry Marks tackled in his reconstruction of the history of infectious diseases and trials of penicillin in the USA. The US NRC had its own Subcommittee on Venereal Disease in 1943 tasked with evaluating penicillin efficacy in infected soldiers. Harry Marks, *The Progress of Experiment: Science and Therapeutic Reform in the United States* (Cambridge: Cambridge University Press, 1997), 108–11.
30. On Antonio Gallego, see María Jesús Santesmases, *Entre Cajal y Ochoa: ciencias biomédicas en la España de Franco* (Madrid: CSIC, 2001), Chap. 4.
31. R. E. Gruber (vice-presidente of Merck & Co.) to Juan Lladó (managing director of the Banco Urquijo), January 14, 1949 (copy). Archivo Antonio Gallego, private collection preserved by his son Roberto Gallego, Alicante (Spain).

32. 'Agreement made and entered into as of the 14th of January, 1949 between Merck & Co., Inc. and Banco Urquijo', contratos 37QC, cajas 5112 y 5113, IDD 1.04. Ministerio de Industria. Sección de Subsecretaría (Asistencia Técnica), AGA, Madrid (in English in the original). I would like to thank Mar Cebrián for her guidance in consulting these files at the AGA.
33. A. Romero de Pablos, 'Regulation and the circulation of knowledge: Penicillin patents in Spain', *Dynamis* 31(2) (2011): 363–83.
34. Acta de la Junta General de 2 de marzo de1949. Libro de Actas de la Real Academia de Farmacia. Quoted by F. Javier Puerto, *Historia de la Real Academia Nacional de Farmacia* (Madrid: Real Academia Nacional de Farmacia, 2012), http://dx.doi.org/ES/monoranf.v0i0.1273, 545–6.
35. Cozzoli, 'Penicillin and the Europeans'.
36. *Boletín Oficial de Estado*, August 11, 1949. Each project approval by the Ministry of Industry and Trade was published separately in successive pages of the *Boletín*: 'Decreto de 17 de junio de 1949 por el que se resuelve el concurso convocado por el decreto de 1 de septiembre de 1948 para fabricar penicilina en España a favor de la proposición presentada conjuntamente por "Consorcio Químico Español SA y Banco Urquijo SA"'; and 'Decreto de 17 de junio de 1949 por el que se resuelve el concurso convocado por el Decerto de 1 de septiembre de 1948 para fabricar penicilina en España, a favor de la proposición presentada por "Industria Española de Antibióticos SA"'.
37. Mar Cebrián, 'Technological Imitation and Economic Growth during the Golden Age in Spain: 1959–1973' (PhD diss., Florence: European University Institute, 2004).
38. 'Nuestro requerimiento al INI', undated but, according to the answering report (see below), ca. 1949; in 'Penicilina y Antibióticos', caja P 44, Archivo Altos Cargos, Joaquín Planell, vicepresidente, Archivo del Instituto Nacional de Industria, Madrid.
39. 'De dirección técnica a gerencia: Solicitud de participación y colaboración en la Empresa Industria Española de Antibióticos SA', 25 de 11 de [19]49, legajo 051, documento nº 202 (1949). Archivo del Instituto Nacional de Industria, Madrid.
40. *Boletín Oficial de Estado*, August 31, 1948. According to CEPA annual reports, the aim of bottling imported penicillin in Spain was to save foreign currency, but the decree, by mentioning the possibility of bottling streptomycin, for which a decision about Spanish production had not yet been made, rather suggests that by 1949 information about the therapeutic effects of this new antibiotic was already widely known and authorities were once more expecting to import the product from abroad for bottling as a first step toward production. See 'Memoria de la Compañía

Española de Penicilinas y Antibióticos, 1952', 2, Fondos Banco Urquijo, Biblioteca del Instituto Juan March de Estudios Sociales, Fundación Juan March. Madrid (hereafter Fondos Banco Urquijo).

41. 'Memoria de la Compañía Española de Penicilinas y Antibióticos, 1949–1950; Fondos Banco Urquijo.
42. *ABC* Sevilla, September 10, 1950, 4–5; same text in *La Vanguardia*, August 20, 1950, 5.
43. On women workers, see T. Kaplan, 'Social movements of women and the public good', in *Dones en movement(s)*, eds. C. Borderías and M. Renom (Barcelona: Icaria, 2008), 19–47.
44. *La Vanguardia*, August 29, 1950, 5.
45. *ABC*, August 16, 1950.
46. M. S. Gewurtz, 'Looking for Jean Daw: Narratives of women and missionary medicine in modern China', in *Figuring It Out. Science, Gender and Visual Culture*, eds. A. B. Shteir and B. Lightman (Hanover, NH: Darmouth College Press, 2006) has been an inspiration for my analysis.
47. *ABC* Sevilla, September 10, 1950, 5.
48. Lina Gálvez, *La Compañía arrendataria de tabacos: 1887–1945: cambio tecnológico y empleo femenino* (Madrid: LID Editorial Empresarial, 2001). See also M. Rentetzi, 'The Tobacco Museum of the city of Kavala', *Technology and Culture* 50(3) (2009): 649–57.
49. Cristina Borderías Mondéjar, *Entre líneas: trabajo e identidad femenina en la España contemporánea: la Compañía Telefónica, 1924–1980* (Barcelona: Icaria, 1993).
50. Maxine Berg, *The Age of Manufactures, 1700–1820: Industry, Innovation and Work in Britain* (London: Routledge, 2005).
51. As is well known, women are still a minority among students in technical and engineering schools.
52. Teresa Gallego Méndez, *Mujer. Falange y franquismo* (Madrid: Taurus, 1983); Margaret Rossiter, *Women Scientists in America: Before Affirmative Action 1940–1972* (Baltimore: Johns Hopkins University Press, 1995); Marga Vicedo, 'The father of ethology and the foster mother of ducks: Konrad Lorenz as expert on motherhood', *Isis* 100 (2009): 263–91.
53. Rosa Medina-Doménech and Alfredo Menéndez-Navarro, 'Cinematic representations of medical technologies in the Spanish official newsreel, 1943–1975', *Public Understanding of Science* 14 (2005): 393–408. On women, feminism and medicine, see Teresa Ortiz, *Medicina, historia y género. 130 años de investigación feminista* (Oviedo: KRK, 2006).
54. C. Sarasúa, and L. Gálvez, eds., *¿Privilegios o eficiencia? Mujeres y hombres en los mercados de trabajo* (Alicante: Publicaciones de la Universidad de Alicante, 2003); C. Segura, 'Mujeres, trabajo y familia en las sociedades industriales', in *La historia de las mujeres: una revisión historiográfica*,

ed. Isabel del Val Valdivieso, Magdalena S. Tomás Pérez, Jesús Dueñas Cepeda, Cristina de la Rosa Cubo (Valladolid: Universidad de Valladolid-AEIHM, 2004).

55. Anna Cabré, 'Cuarto aproximaciones explicativas a las tendencias de nupcialidad y fecundidad', in *La constitución familiar en España*, ed. A. Cabré (Madrid: Fundación BBVA, 2007); Margarita Delgado, 'Cambios recientes en el proceso de formación de la familia', *Revista Española de Investigaciones Sociológicas* 64 (1993): 125–54; Luis Garrido, *Las dos biografías de la mujer en España* (Madrid: Instituto de la Mujer, 1993); M. J. Santesmases, *Mujeres científicas en España: profesionalización y modernización social* (Madrid: Instituto de la Mujer, 2000).
56. *ABC*, December 31, 1950.
57. I am grateful to the archivist Thore Grimm at Schering for showing me these photographs that helped put my argument into perspective.
58. Anonymous, *Revisión de los conocimientos actuales sobre la Penicilina*, unpaginated between 48 and 49 and between 56 and 57.
59. See two photographs of women at the Parke-Davis manufacturing plant in Scott Podolsky, *The Antibiotic Era: Reform, Resitance, and the Pursuit of a Rational Theraputics* (Baltimore: Johns Hopkins University Press, 2015), 24.
60. *La Vanguardia*, July 20, 1950, 5.
61. 'Memoria de la Compañía Española de Penicilinas y Antibióticos, 1952', 1. Fondos Banco Urquijo. This was the same amount as was produced in the USA in January 1944; see Hobby, *Meeting the Challenge*, 196.
62. I owe Sagrario Mochales for this testimony: Interview with the author, Madrid, 2012.
63. Santesmases, *Mujeres científicas en España*, Chap. 3.
64. *ABC*, July 18, 1951; *ABC*, June 30, 1957.
65. *ABC* Sevilla, September, 1950, 5.
66. On retrieving women, see Montserrat Cabré, 'Women or healers? Household practices and the categories of health care in late medieval Iberia', *Bulletin of the History of Medicine* 82 (2008): 18–51.
67. The increasing demand for penicillin detected by CEPA in 1950 was connected in their annual reports to a detected demand for streptomycin. See 'Memoria de la Compañía Española de Penicilinas y Antibióticos, 1952', 3. Fondos Banco Urquijo.
68. 'Memoria de la Compañía Española de Penicilinas y Antibióticos, 1952', 3–4. Fondos Banco Urquijo. Servicio de Estudios.
69. Joan Scott, 'Gender a useful category in historical analysis', *American Historical Revue* 91 (1986): 1053–75; Ludmilla Jordanova, 'Gender and the historiography of science', *British Journal for the History of Science* 26 (1993): 469–83; Ilana Löwy and H. Rouch, 'Genèse et development

du genre: les sciences et les origins de la distinction entre sexe et genre', *Cahiers du Genre* 34 (2003): 5–16.

70. Inspired by Ludmila J. Jordanova, 'Natural facts: A historical perspective on science and sexuality', in *Nature, Culture and Gender*, eds. C. P. MacCormack and M. Strathern (Cambridge: Cambridge University Press, 1992), 42–69, 57.
71. See the 'Memorias de la Compañía Española de Penicilinas y Antibióticos' from the 1950s and early 1960s. Fondos Banco Urquijo. Servicio de Estudios.
72. Agata Ignaciuk, 'Publicidad farmacéutica y género: Anuncios de la píldora en España (1964–1985)', *Medicina e Historia* 1 (2014): 4–21. Jean-Paul Gaudillière and Ulrike Thoms, 'Pharmaceutical firms and the construction of drug markets: From branding to scientific marketing', *History and Technology* 29(2) (2013): 105–15.
73. Cazorla, *Fear and Progress*.

CHAPTER 4

Smuggling: The Management of Scarcity and Trade of Penicillin as a Post-War Commodity

Reflecting on archival material relating to smuggling—documented in personal memoirs for the 1940s and in public archives for the early 1950s—alongside the relevant historiography, enables an analysis of the cultures and social life of penicillin, and its relation to other scarce commodities of the time, such as tobacco and nylons. The practices which surrounded obtaining the drug demonstrate the relationship between official secrecy and public knowledge, and its embeddedness in the political strategies of Franco's dictatorship. Penicillin will also be discussed in this chapter as both a border-crossing material and symbolic object, representing the ongoing therapeutic revolution. The historiography on corruption suggests the illegal trade in penicillin might also be linked to the 'business of power', when, for example, gasoline supplied for official use was being sold on the black market.[1]

During the night of October 17, 1950, close to the village of San Fernando, near Cádiz on the south Atlantic coast of Spain, members of the Guardia Civil spotted a number of men with loaded *caballerías menores* (donkeys) in the *salina Santa Cruz* (open air salt mine) and ordered them to stop. The men were salt miners transporting tobacco and penicillin. Suspecting there would be more parcels in the surrounding areas, the police discovered additional amounts of both penicillin and tobacco at the *Santa Cruz* and at another salt mine close by, *salina la Teresa*. The Guardia Civil confiscated both donkeys and merchandise, and the riders were interrogated. The salt miners reported that the merchandise belonged to

M. J. Santesmases, *The Circulation of Penicillin in Spain*, Medicine and Biomedical Sciences in Modern History, https://doi.org/10.1007/978-3-319-69718-5_4

a man from Tangier, in the north of Morocco, who had negotiated with one of them for the transport of some packages from one salt mine to the other. This man had hired six of his salt-mining workmates who needed the money—salaries were extremely low—and they had used some of the mining company's donkeys to bring the goods to a transportation vehicle. Every miner was interviewed by the police that night. All claimed to be unaware of the final destination, and believed the mine owner did not know about the deal. The contraband consisted of nearly 39,730 packets of cigarettes and 5600 flasks of penicillin G, each containing 200,000 units. The penicillin, handed over to the Customs Administration in Cádiz, makes no further appearance in the report. The tobacco, however, was considered a major seizure, and the case was successfully tried almost four years later at the Cádiz Court of Smuggling and Fraud. Only the miner who had recruited the others—a 'repeat offender'—and the 'man from Tangier', were declared guilty of smuggling.[2]

This narration is included in one of the earliest and most detailed reports I have found in the Tribunal de Contrabando y Defraudación (Spanish Court of Smuggling and Fraud) archives which involved penicillin. It is also one of the few large cases of smuggling, according to the Court's classification. It includes a set of unknowns in the declarations of those involved: they apparently had no knowledge of who had brought the merchandise to the place they found it, what the merchandise was, where it had come from, or where it was supposed to be going. Lightweight drug flasks could be transported easily from one place to another in small amounts, and penicillin, being in great demand, proved a lucrative business for dealers and enabled the salt miners in Cádiz to earn extra money. Not only were the major dealers never found, they were not even referred to in the report. The salt-mining firms were declared not guilty; their lawyer argued they were unaware of any smuggling.[3] Although illegally brought to the place it was intercepted by the Guardia Civil, the smuggled penicillin was clearly identified as being manufactured, bottled and marketed by Merck: according to labels the drug had been produced in the USA.

Penicillin, along with tobacco, was being detected and confiscated when transported from one place to another: while travelling. It circulated in large, lightweight cardboard boxes, which at that time contained penicillin in glass vials or ampoules. The yellow powder—as penicillin was often referred to in early times—was a dry extract sealed in blown glass, and could only be used once as the drug became inactive in open air. Once industrially manufactured in its pure form, penicillin turned white and could be stored in small bottles with a rubber top covered in aluminium; a central circle could be extracted and the rubber punctured with a syringe. Oil, beeswax or distilled water would be added and the resulting suspension extracted for intramuscular or intravenous administration.[4] The lightness of the vials and the boxes containing them facilitated transport and the ability to be hidden in many secret locations. The box size, along with its weight, would have made it recognisable, however, to a keen observer: but observational expertise regarding penicillin in 1940s Spain would have been as limited as the drug itself. Its rarity added to its symbolic value as part of the post-Second World War culture of science, technology and medicine it was contributing to. Box labels revealed the drug's origins but remained silent about the agents of transportation and delivery. Packaging has played its part in this reconstruction as it has in the histories of other epistemic objects[5]: as a moving agent and representative of its contents, the illegally circulating and conveniently lightweight miracle drug.

Smuggling, the black market and illegal trade are terms associated with both scarcity and the policy-making of governing scarcity. The economic system that stabilised at the end of the Second World War was composed of both legal and illegal trade. The poverty, hunger and disease of post-war devastation created a demand for food and also for medicinal drugs, but national governments lacked resources and the old smuggling practice of crossing national borders without permission and enabling goods to circulate maintained its popularity. As Oscar Rodríguez Barreira has stated, the black market created a popular way of life, as a strategy for subsistence and a practice of daily survival. At the same time, illegal trade assisted the creation of the dictatorship's corrupt economic and political system. Franco's bureaucratic and social elites benefited from the black market; this 'weapon of the weak', which enabled many to survive, also helped sustain the regime. My aim in this chapter is to include penicillin in this regime, by depicting the social, legal and illegal trajectories of penicillin, a post-war commodity that, alongside other medical drugs and health policies, participated in the emergence of a political economy.[6]

Food Rationing and *Estraperlo*: A Pre-Existing Practice

As the Civil War ended, before any information on penicillin could have been received, food was the priority in the regulations and political discourse of Franco's regime. The insufficient food available in Spain was centrally distributed through ration cards. Administered by the Comisaría General de Abastos y Transportes (General Office for Supply and Transport), rationing came into force in 1939 and lasted until 1952. Initially, with the stated intention of equal distribution among the entire population, a basic, minimal ration was determined for all: this was later adjusted for age and gender-based criteria. The combination of a complex supply network, rigid system and the regime's toleration of corruption led to the immediate emergence of a black market, enabling the wealthy to access extra food in a society struggling to supply basic nourishment. Illicit white bread, olive oil and flour could be obtained from agriculture and small-scale manufacturers, or from the rations of families needing to buy other commodities.[7] Thus, through a combination of legal and illegal means, the Spanish people attempted 'the management of hunger'.[8]

With the compulsory sale of farm produce to a state monopoly at fixed prices, the black market of foodstuffs thrived. In 1940, Franco reacted by creating a body to prosecute illegal trade, the Fiscalía de Tasas. Each province has its own Fiscalía Provincial court, with recourse to the Fiscalía Superior as a court of appeal. In small villages the fear of being visited by Fiscalía agents prompted an increase in black market prices: the bigger the apprehension the more expensive goods became, and further regulations soon increased the harshness of penalties.[9] A local archive of Fiscalía de Tasas documents has been preserved and provides insightful historical reconstructions of the smuggling and its punishments during Franco's regime. The creation of these courts did not eliminate the black market, however, as such illegal practices were tolerated by the authorities. According to historian of agriculture, Carlos Barciela, fifty per cent of the wheat consumed during the 1940s passed through the black market. A large part of the economy was therefore evading state control, and contributing to social inequality, leading Barciela to consider Franco's political economy responsible for the stagnation of Spanish development, hampered as it was by the lack of open, legal trade.

In the immediate post-Civil War era, therefore, the Spanish black market was not only a way of overcoming scarcity but the conventional way of conducting everyday 'business'. Certainly a blind eye was turned to the small amounts of food and other goods that crossed the border between France and Spain.[10] By the end of the Second World War new commodities had also began to circulate illegally. Nylons and penicillin were both results of the policies promoting research and industrial production during the Second World War in the UK and USA.[11] These new products came to Spain illegally, avoiding customs and police controls by marine transportation—the Iberian Peninsula had many active seaports—and through the Pyrenees. During the 1930s and until the end of the Second World War, guides who knew every possible path through the mountain chain separating Spain and France had assisted in the Jewish migration to Portugal and from there to America. The Pyrenees and other mountainous areas in Spain also provided protection for the *maquis*, the resistance movement that remained hidden during the post-war years. These mountains—as geographies to be crossed—participated in the smuggling of new medical drugs, commodities and tobacco. The illegal distribution of these new goods was certainly not intended to overcome basic food shortages. Penicillin, however, was becoming a basic need. Its success at curing infections had created a new market focused on health, human bodies and infective, fatal diseases. So the Pyrenees became a border, as the seaports were, through which not only people but products—medical drugs, and many other commodities—arrived.

Early Circulation, from the Front

As discussed in Chap. 1, the first reports of penicillin use in Spain had appeared in March 1944. This penicillin had come through Gibraltar, the British harbour on the south coast of Spain, from an 'African port occupied by the Allies'.[12] The clinician treating the patient in hospital 'claimed to know nothing of the deals made so that this famous pharmaceutical could be obtained'; both at the bedside and on the black market, penicillin worked efficiently thanks to its non-regulated availability. Until penicillin was available from pharmacies, it was subject to intense state bureaucracy: the first parcels from the USA were placed at the service of the political bureaucracy. Only from 1950 onwards did penicillin become a product sold to and for everyone. Both bureaucracy and scarcity became the basis upon which smuggling and illegal trade developed. Coffee and cigarettes from abroad had been the most widely circulated goods in earlier times, while basic foodstuffs were traded on the black market throughout the decade in which rationing was in force.[13]

Penicillin was not regarded to be a basic need at the time rationing was implemented, and was therefore never included—neither were many other pharmaceuticals of the time—but its increasing socio-medical relevance and renown were rapidly transforming the drug into a 'basic' live-saving commodity. Penicillin had also been a rare and celebrated good in the USA during the Second World War, and was vehemently demanded by the ill and their families in every country until it became available in sufficient amounts.[14]

Exchanges between Allied forces and the German army were one of the earliest paths for news of this miraculous yellow powder. In places where Allied Army physicians had penicillin, German POWs had the chance to witness this phenomenon. British penicillin allocated to the African front could well have been the mysterious origin of the drug used to treat the Spanish mining engineer in March 1944. Without any further details, the report suggested either Oran or Algiers as the drug's origin, both well-known sources of contraband tobacco at that time.[15] This unknown or secret origin suggests the penicillin was illegally obtained, or at least that it was not registered as a result of any agreement or humanitarian donation. The same day, according to news reports, another sample, preserved in ice and apparently obtained by legal means from Brazil, was brought by the Portuguese ambassador himself through the Brazilian Embassy to treat a young girl. Her father and aunt appeared in a photograph published by a Madrid daily with the parcel containing twelve 5 ml ampoules of caramel-coloured oil.[16] This penicillin sample proved to be extremely expensive, calculated at around $15,000 on the basis of the four consecutive flights required for its transportation, apparently from the Oswaldo Cruz Institute near Rio de Janeiro.[17] After some promising improvement, the girl's condition deteriorated and an attempt was made to retrieve penicillin from her urine, a practice previously performed in Oxford during the first testing of minute samples by Ethel Florey, who had established penicillin's topical action by 1942.[18] Treatment with penicillin was not always successful; the wife of a well-known bullfighter had died that April following treatment, at the age of thirty-two.[19] Around the same time, a clinician in Barcelona received a parcel 'from a Cuban laboratory with two million Oxford units', which he used to treat two patients the following month without any definitive results.[20] Samples of the mould *Penicillium notatum* were also circulating in the city of Delft in the Netherlands under Nazi occupation,[21] and a medical researcher in Barcelona who received a mould sample from the

Lister Institute in London reported its properties in laboratory animals at the Academy of Medicine in April 1944.[22] Circulation of the mould for culturing was widespread. Fleming was sending samples to laboratories around the world by that time, where it would be purified, and the industrially manufactured commodity eventually fully controlled by laws and national customs (see Chap. 3).

As stated in Chap. 1, the clinician Carlos Jiménez Diaz had been diagnosed as suffering pneumonia while on holiday in the summer of 1944, and was successfully treated using penicillin. The drug had been bought at a downtown public house in Madrid. Having refrigerators, pubs and other food and drink establishments were convenient places to store penicillin while waiting for clients. Refrigerators were rare in homes, where perishables and freshly preserved penicillin would be kept in metal boxes. The drug was therefore transported in ice bags to Jiménez Díaz's bedside every day of his treatment. This was at a time when the Spanish health authorities claimed to have no penicillin at all.[23]

Gregorio Marañón, another highly respected clinician at the time, published a report of a female patient diagnosed with staphylococcal meningitis in 1945: results from the penicillin he had 'managed' to obtain were astonishing, as the infection had appeared to be 'inexorably deadly'.[24] 'In some years', he added, this will be 'an ordinary event' but on that occasion he was extremely moved by the effects. While using what he regarded as a small amount of penicillin—'only one million units'—some 'accidents' also occurred during the treatment: the patient experienced pain in her legs and aphasia, but recovered fully from both. The drug may not have been entirely pure, as Marañón reflected on whether these side effects could have resulted from the drug itself or its impurities.

Another report on the potentiality of the new drug was published by the physician Federico Tello in 1946, who reported promising results from treating mycosis of the lung with penicillin; the condition had previously carried an almost hopeless prognosis.[25]

Following the Normandy landings on D-Day, the public renown of penicillin increased dramatically as the US army distributed the drug among troops on the European front. Political leadership and an agenda promoting the industrial development of health care travelled attached to every ampoule, laying the foundations for penicillin production in every European country. But penicillin had circulated earlier than this date, at the origins of the end of the Second World War.

The Arrival of Penicillin from the USA

The parcel of the antimicrobial that arrived at Madrid airport from the USA via Lisbon in September 1944 was the first of a series of deliveries, 'despite the huge demand for civil uses and on the war front', resulting from an agreement between the Ministry of Industry, health officials and the US Embassy, according to the Madrid daily, *ABC*, and Barcelona daily, *La Vanguardia*.[26] As Chair of the US Committee on Chemotherapy, Chester Keefer had recently published on the promising results obtained with penicillin. Based on his 'personal observation of a considerable number of patients', Keefer believed the 'non-toxic' drug 'would save many lives and reduce many days of acute and chronic illness'.[27] Over 2700 hospitals in the USA received monthly supplies of the drug that year despite strict controls on distribution; it was not until December 1944 that US production met military requirements and civilian needs could be fully addressed. According to Gladys Hobby's historical reconstruction, the US War Production Board was allowing the sale of penicillin 'through normal trade channels' by early 1945.[28] The scarcity in Spain endured far longer, however, fuelling illegal trade.

The first batches of penicillin received in September 1944 were put under the control of the Comité Nacional, as discussed in Chap. 1. According to a news report the penicillin originated from ten of the 20 US plants that manufactured the drug. Each ampoule cost approximately 30 pesetas, making a course of treatment around 300 pesetas, an extremely high price for a poverty-stricken nation. Restrictions were placed on its use to avoid price increases and priority was decided according to patient need. The impressive results of 2000 treatments were reported to the Comité Nacional, and to the US Embassy, which proceeded to send them to the US Commission collating results.[29] Penicillin distribution was strictly controlled by the Spanish state, which apparently paid for subsequent batches from 1946 onward.[30] This controlled distribution created a rationing system in which penicillin was sold at a fixed price. Rationing, scarcity and bureaucracy were the basis upon which a black market of the new drug emerged, controlled and regulated by the same norms as food rationing and its illegal trade.[31]

The Franco government fully embedded the drug in state bureaucracy, and strict rules for the distribution of the drug were approved in November 1944.[32] Chaired by clinician and medical researcher, Carlos Jiménez Díaz, the Comité Nacional consisted of both clinicians and health administrators, among them Jiménez Díaz's former colleague the endocrinologist Eduardo Ortiz de Landázuri, bacteriologist Gerardo Clavero, dermatologist Enrique Sáinz de Aja and pharmacist Nazario Díaz.[33]

Immediately following arrival at Madrid airport, the humanitarian parcels were placed in a refrigerator to await import procedures.[34] Importing of medical drugs was severely restricted: approval was required from both health and industrial officials, and only those drugs that had 'therapeutic relevancy' for public health would be considered. Foreign firms authorised to manufacture medicinal drugs before the Civil War had to obtain 'special permission' from the government to continue. State regulation of pharmaceutical production and trade appears to have been more oriented to promoting national production than assisting any public health policy: no guarantee was made that the general population would be provided with pharmaceuticals.[35]

Once supplies of penicillin began to arrive, the new drug reinforced the dictatorship's power. According to the regulations on penicillin distribution approved by Franco in November 1944, a detailed clinical report including analysis and diagnostic tests of every patient who applied for the drug had to be supplied. Once the Comité Nacional had approved penicillin treatment and the supply had been authorised by health officials, physicians were required to provide a daily treatment report, which the Comité Nacional could verify by visiting the patient at any time to check data and doses. Surplus units had to be returned for use in charitable medical centres at the discretion of the General Director of Health, which suggests the patient would have to pay for the penicillin.[36] The Comité Nacional would also determine the cost of the drug.[37] The main functions of the Comité Nacional therefore evoke those of ration cards: the management of scarcity as a policy of adjusting supply and demand, within a rhetoric that claimed those in most need would receive the few units available, the stated intention of early bureaucratic control. Some enduring fortunes created in Spain at that time, however, were created by corruption, by taking full advantage of rampant poverty and hunger, as agents in the black market. In December 1944, after a meeting of the Consejo Nacional de Sanidad at which it was declared the health of the general population 'could not be better', news of the supply of penicillin from the USA was released.[38] Such a comment on a supposedly healthy population contributed to a permanent suspicion of official statements, so few among the general population would have given the stated intentions of penicillin regulations credence.[39]

A Silent History

By the end of the Second World War, when US and, soon after, British companies were producing enough penicillin to cover increasing demand, the drug began to circulate more widely. Spanish authorities, however, maintained tight control of the apparently scarce amount received from the USA, and the relatively limited measures taken against the black market suggest its existence was largely tolerated.[40]

From the arrival of the first humanitarian batch onwards, news about penicillin, abstracts of publications about its manufacture and use, and medical reports on the treatment of infections were published in Spanish medical journals. Some articles reporting successful treatments mentioned the unknown origins of penicillin obtained by patients, thus acknowledging both its scarcity and the existence of illegal trade in the renowned drug. Scarcity contributed to penicillin's legendary status: it was not cheap and not everyone who needed the drug was able to obtain supplies. Christiansen (2005) has argued that in regard to the black market of food during the 1940s and early 1950s, this was the norm, not only in Spain but throughout Europe, in both democratic nations and dictatorships.[41] This suggests that for penicillin, the emergence of a black market resulted from the combination of a state monopoly purchasing the drug at fixed prices and a consumer rationing system. Demand outstripped supply in Spain until domestic production commenced in the mid-1950s (see Chap. 3).

A black market in this penicillin and, slightly later, in streptomycin—the effects which against tuberculosis were even more astonishing—is often mentioned in both fact and fiction from this time. 'My first experience with penicillin was the closest to a miracle I'd ever seen', declared a village doctor in northern Spain when remembering his medical practice at the end of the 1940s: 'By that time penicillin was not being sold, it has to be obtained in the black market, from smuggling.' A young patient of his presented with pneumonia, and following unsuccessful treatment with sulphonamides, deteriorated so severely the local priest was asked to perform the last rites. A boy with a motorbike was sent to the nearby port of Gijón—the doctor knew exactly where the drug was sold—where it was possible to find penicillin, 'fully guaranteed' to work. An hour later the doctor was presented with 200,000 Oxford units of a drug he had never previously used. Having prepared a solution, he injected 18,000 units every three hours, and stored the flask in ice to maintain its activity.[42] Penicillin was a business opportunity for those involved at any level. Scarcity made access difficult, but increased financial rewards. The controllers of this life-giving medicine, as Robert Bud has described penicillin, received impassioned and desperate requests, the majority impossible to fulfil through legal procedures.[43] During the long post-war, clinical confidence in the new drugs positioned health care systems and the pharmaceutical industry as major players in the reconstruction of Western societies and contributed to the promise of a modern, healthy world.[44]

As discussed in Chap. 3, the Spanish government became fully aware of the growing demand for penicillin in September 1946, when a report recommended manufacture of the drug in Spain.[45] Authorities at the Ministry of Industry discussed the proposal in 1947 and 1948, at a time when No-Do footage of Alexander Fleming's visit to Barcelona, Seville, Córdoba, Madrid and Jerez was depicting the fervour with which the British researcher had been received. As well as being powerful propaganda for penicillin, the films provided strong support to the manufacturing proposal already on the table at the Ministry of Industry.

On September 1, 1948, the Spanish government took control of penicillin manufacturing, including it in their industrial priorities, and in August the following year, authorised two firms to manufacture penicillin in Spain. The firms signed contracts with US firms—one with Merck, the other with Schenley—enabling them to use their manufacturing methods under license. Both firms were private, but the state controlled, by law, their financial and foreign relations activity.

Penicillin was initially received in bulk from the USA, and by the end of 1950 one of the companies alone had bottled half a million flasks of the drug.[46] This penicillin was in the pharmacies and hospitals the following year. The 1940s were over, and autarchy and ration cards would soon follow; smugglers, who until then had been at least partly tolerated, would now be prosecuted. A transition had taken place: from a scarce remedy in the 1940s, penicillin had now become a cultural icon associated with industrial modernisation and prosperity. Modernity at that time, however, was also compatible with smuggling, illegal trade and, more broadly, the permissive practices of authorities toward this illegal trade in times of scarcity. Penicillin meant business as well as therapy. Its success in treating increasing kinds of infections kept the miracle of penicillin alive.

Smuggling: Denouncing and Punishing Illegal Trade

That the general public were aware of the smuggling of penicillin and, soon after, other antibiotics, is illustrated by the Spanish film, *Mercado prohibido* (Forbidden Market), directed by Xavier Setó and first shown in 1952. A portrait of the post-war climate in Spain, it centres on the leader of a group of illegal antibiotic dealers who needs chloramphenicol for his own son, suffering from typhoid fever. In order to save his son's life, he begs for help from those he has deceived and betrayed and, before he can visit his son and administer the medicine, is finally arrested.[47] The film was heavily influenced by *The Third Man*, a 1949 British film noir directed by Carol Reed, with a screenplay adapted by Graham Greene from his own novella. The plot involves the diluting of stolen penicillin for sale on the black market in post-war Vienna. *Mercado prohibido* uses the same tilted shots and marked, contrasting shadows of black and white. The Spanish version focuses on penicillin, streptomycin and chloromycetin—Chloramphenicol by Pfizer—and mentions the promises of Spanish penicillin production from 'one of the biggest' factories in Europe, that would eventually vanquish the black market altogether.

In 1952, under the Law of Smuggling and Fraud, Franco created the Tribunales Provinciales de Contrabando y Defrudación (Provincial Court of Smuggling and Fraud), and the *Tribunal Superior* to deal with appeals against provincial court decisions. According to the law's preface this was due to the 'extension of smuggling and fraud' and the 'impunity' of such practices, thus suggesting the local Fiscalías de Tasas had not been able to operate effectively: creating a specific court, it was hoped, would enable more efficient prosecutions of both smuggling and fraud.[48]

It was during the first months of 1954 that CEPA detected 'a great disturbance' in the penicillin market. Large amounts of foreign penicillin were reaching Spain through 'clearly clandestine ways'. This produced a *decalage* between production and sales that was severely undermining the profits of both CEPA and Antibióticos SA. Surplus penicillin manufactured in the USA entered the international market during the early 1950s, according to CEPA annual reports, which not only depressed prices as Spanish laboratories bought penicillin at 'dumping' prices, but put illegal—though bottled and at times labelled—penicillin into circulation, which, Spanish health authorities argued, meant quality levels could not be guaranteed.[49] These large amounts documented in CEPA archives, shared public and private space with smaller quantities of illegal penicillin at a time when the two factories were producing enough penicillin to cover national demands and had already begun manufacturing streptomycin.

I have reconstructed the events related here from the archives of the Courts of Smuggling and Fraud. I have not found any reports of penicillin in bulk, only bottled, which suggests many batches remained undetected and successfully entered the market. Prosecutions for possession of smaller batches had been made by the late 1940s, before the new law came into force. The early distribution of penicillin involved time and geography, a particular chronology of post-Second World War geopolitics. Penicillin practices and experiences belong to this specific political period, in a Spain governed by a dictatorship that tolerated the smuggling and illegal trade of other products. Some of these, like penicillin, were manufactured in countries bordering Spain, in a wider neighbourhood that was both geographical and political: the West.[50]

Although official supplies of penicillin were being established by 1950, smuggled penicillin was also available. Police would often intercept smugglers on moonless nights, having usually received a *soplo*, a tip-off; according to court reports based on statements from those involved, goods would be brought to a beach or near the shore on a cargo ship sailing under a foreign flag, anchored off the Spanish coast to take on provisions for its onward journey. Those who found smuggled goods received a reward; discoveries encouraged with financial incentives. The intended final destination of confiscated contraband was not made clear in court reports. Fines and prison sentences were meted out to those found transporting or possessing penicillin without permission, and from 1950 on, confiscated penicillin would be sold at auction. Although the products smuggled into Spain most often at that time were coffee and cigarettes, other commodities, such as nylons, nail polish and medicines joined the basic foodstuff that had fuelled the black market since the 1940s.[51] The local Court of Smuggling and Fraud received military police reports of any smuggling activity and subsequent confiscation of penicillin. Although not as popular as coffee, cigarettes and nail polish, or at least less often intercepted, the drug circulated widely. It was kept in homes and pubs, and transported on trucks, coaches and in baggage deposited on trains, to be collected by, among others, a woman waiting at the train station or carried by a man in a bag down a town street near the seaside. In the midst of an extremely vigilant society—as Spain was in the early 1950s—smuggling was severely prosecuted once the distribution of legal penicillin began.

As narrated at the beginning of this chapter, the amount of penicillin seized at San Fernando in 1950 was substantial. In the years to come, the Guardia Civil would intercept various quantities of flasks, from ten to hundreds. In May 1953, another large batch was intercepted: 22,000 flasks of Merck penicillin and 5600 of dihydrostreptomycin—along with nylons, toothbrushes, flashlights, lighters, nail polish, sugar and cigarettes—were found on a Danish boat anchored 500 metres west of Punta Antequera on the Canary Island of Tenerife. The person in charge was from Tangier and stated that the boat was on its way to Monrovia.[52]

The earliest smuggling case I have found is dated December 1949, when three flasks of 600,000 units of penicillin, 175 grams of streptomycin, nylons and plastics were found by the police in a private apartment. The owner handed the parcels over voluntarily, claiming to have no idea who had brought the goods there or where any invoices might be found, after police threatened to register the whole house.[53] At the train station in Alcázar de San Juan, a village in the province of Ciudad Real, central Spain, the police found 'varied amounts of penicillin' in a passenger's luggage. The man admitted to buying the drug from a man in Algeciras and attempting to take it to Valencia.[54] In August 1951, 2.6 miles of the coast of Cabo Salinas on the Balearic Island of Palma de Mallorca, a British-flagged shore boat was stopped and registered by the military coast guard. Twenty-five boxes of penicillin and 329 bags of sugar were discovered, being transported from Gibraltar to Marseille in France, via Oran.[55] In a garage in Seville, 50 million units of penicillin were found in October 1952 by the Servicio de Vigilancia y Represión del Contrabando (smuggling police), along with 162 packs of tobacco and 563 boxes of pencils.[56] In January 1953, 1416 flasks of penicillin were found on a train at Atocha station, Madrid, packaged under the description of 'electrical material'.[57]

In Córdoba in March 1953, police found a post office employee with one parcel containing '20 million' units of penicillin (no number of flasks specified) and another of 30 million, which the employee said his family needed.[58] Also in March 1953, when police visited a Bilbao pharmacy, the pharmacist's daughter voluntarily handed over ten boxes of penicillin and 50 of hydrostreptomycin, along with a number of flu vaccines. The police registered the pharmacy but found no other illegal batches, the daughter admitting buying the drugs herself to meet the urgent demands she was unable to with legal penicillin: according to the report, she was unaware her actions were illegal.[59] In May 1953, a parcel containing

200 flasks of penicillin G were found in San Sebastián on the north coast of Spain, in the warehouse of a transport company. The parcel's label described its contents as stationery and office materials. The person to whom it was addressed declared his innocence, claiming he had not requested any penicillin.[60] In June 1953, three suitcases with 565 flasks of Merck penicillin were discovered in Seville, on a bus from Algeciras.[61]

In July 1953, penicillin and other commodities—including whiskey, gin and cocoa—were found in a warehouse at Barcelona harbour. The penicillin had apparently been brought to treat a driver working at the warehouse, and had arrived from the Andalusian town of Málaga, in southern Spain.[62] In October 1953, a truck was intercepted in the south of the Canary Island, Gran Canaria, which, in addition to sugar, blond cigarettes and margarine, was carrying 74 boxes containing a total of 29,000 flasks of foreign penicillin. In November of the same year, at the bus station in Granada, Andalusia, a traveller's luggage was found to contain 400 flasks of foreign penicillin. The man stated the drug was for his two brothers, one intending to distribute it among his friends, the other was a pharmacist. The flasks were confiscated and a year later the court fined him double the cost of the penicillin. Later the same month, 280 flasks were found in another piece of luggage at the same bus station, apparently coming from Málaga.[63] In December, at a transport agency in the Basque province of Guipúzcoa, northern Spain, a parcel was found containing 100 flasks of penicillin. The woman who had sent the parcel from Barcelona, being unable to prove its legality, was subsequently fined.

In April 1954, again at the bus station in Granada, a driver gave the station chief a parcel containing 15 flasks of penicillin, with a third person stating it had been sent from Málaga to treat a sick aunt. In May of the same year, the local court of smuggling and fraud in Málaga reported the arrests of two men by the military police: one had been carrying a large box containing 225 flasks of penicillin, the other, a bag of 250 flasks of the drug. Neither man had an import license.[64]

All these penicillin confiscations took place at a time when the drug was being legally distributed in Spain, from bulk importation and bottling, as previously described, and from the beginnings of domestic manufacturing. This suggests either there was insufficient legal penicillin, or the drug could be obtained from unofficial sources at lower cost due to a surplus on foreign markets. The available sources do not provide clear information about prices and availability, however, only the amounts of the drug confiscated. Legal production, with strictly controlled prices and a rigid import system, was therefore sharing the market with illegally traded penicillin. The smuggling of food and other necessary products appears to have been at least partially accepted, despite the creation of specialist courts. Legality was challenged every day by this double-faced practice of regulating while participating in corruption.[65] The social biography of penicillin reveals a wide practice impossible to quantify with the archival resources available, and the existence of which has been covered by the historiography and preserved by public and private recollections. As a social practice of concealment, the black market belongs to the history of post-war goods and commodities, penicillin included.[66]

The Long Post-War Period of Legal Stabilisation

While it remained scarce and illegal, penicillin was expensive. Once the government had promoted and authorised manufacture in Spain, and when penicillin stocks reached a surplus, illegal batches of the drug from abroad circulated through Spain and brought prices down. By making it cheaper, the black market participated in the norms and rules of the legal economies of capitalism at the time: the more availability, the lower the price. Powerful economic policies and bureaucratic life therefore governed not only the legal, but also the illegal trade of penicillin. And although the trade might be prohibited, the penicillin itself was legal, branded and from a known source: the name of pharmaceutical producers was usually included in smuggling reports.

The drug was embedded in the politics of its time: in war and peace, as a mythic marvel and a powerful, therapeutic reality, its price regulated by an international market that was, for a time, as powerful as the norms imposed by Franco's dictatorship. Legal penicillin, strictly controlled, initially by the USA and then the Comité Nacional, finally joined the market: first bottled, and a short time later, manufactured in Spain, while illegal trade in the drug played a part in its social life. As with tobacco and other products, the duopoly of penicillin manufacturers would perhaps benefit from the high price of the legal, taxed product. Alien to the discipline imposed by the Franco government, illegally distributed penicillin—on which tax had not been paid—was not regulated, but according to public records, recollections and memoirs, circulated widely during the 1940s, before importation of US penicillin and the start of domestic manufacture in 1952. As both the legal and the illegal coexisted, penicillin played its part in the economic reconstruction of the post-war period, having already contributed to the Allied victory.

Escaping from the structures of professed modernity, penicillin smuggling as a post-war practice and the experiences of both sellers and buyers resisted standardisation and bureaucracy. Attached to the early distribution and strictly controlled circulation of the rare commodity penicillin was during the Second World War, was a secret, whose uncontrolled dissemination took place through the troops at the front, along political geographies, and over time.[67] State bureaucracies drew legal borders for its circulation, as Chester Keefer and other US medical officials prioritised the treatment of troops at the front. If we take into account factual and fictional accounts, as well as Barciela's conclusion on wheat trade during the same period, the meagre batches confiscated by police in Spain during the late 1940s were nothing compared to the actual network of dealers and customers involved.

A political economy of frontiers played a part. The emergence of Allied power during the immediate post-war period, headed by the USA, was preceded by the early distribution of humanitarian batches. Scarcity remained until penicillin began to be bottled in 1950 in Madrid and León by the two private firms approved by a state policy promoting industrial production during the 1940s, the decade of autarchy. And even after this, the high legal price, compared to abroad, would have contributed to keeping smuggling alive, providing foreign penicillin at a cheaper price.

As a commodity that crossed borders, in all its transits between the law and hidden trade, between countries and social domains—between war fronts and from a war front to an urban site where it was to be sold—penicillin reveals practices in the early days of a promising prosperity, in sight by the 1950s. These transits were permanent tests of a society based on taxes and exchanges, law and bureaucracy, control and discipline. By following penicillin across geographical, political and professional frontiers, we see the drug as a border-crossing commodity that challenged yet also stabilised norms and laws: illegal penicillin contributed to its renown while placing a political-economic system that was validating bureaucracy and control of the same substance with only partial success in jeopardy. The official route fit with the economic interests of Cold War policies: jurisdiction combined with a free market, through the fissures of which penicillin escaped police control. Legally manufactured, it moved into the illegal space of smuggling and fraud when already an object belonging to an international network of trade.

The Clandestine Circulation of a Healing Commodity

Watching patients recover when treated with the first units of penicillin available for testing, physicians were moved by the impressive results, and had great hopes that a time would come when the drug would be available for everyone in need. Until the time officialdom could bring this about, the black market would provide.

Various types of transportation were used to distribute illegal penicillin, including boats, donkeys, coaches and luggage of any kind. The amount of flasks could range from ten, perhaps for the treatment of a relative, to thousands for a medium-sized business. Penicillin was sought for, found and confiscated; those responsible judged and fined. The public life of the drug crossed any boundary, despite police surveillance during the 1950s being unremitting and, at times, efficient. I could not find any archival material on the confiscation of penicillin before 1949, when it had been scarcest and increasingly in demand. The 1940s was the decade of food rationing cards, the most repressive period of the Franco dictatorship, when corruption and the black market prevailed. Scarcity has been the arena in which the life of illegal penicillin has been both most familiar and most poorly documented. In keeping with the historiography of the Spanish black market for wheat during the same decade, it could be suggested that smuggling was to some extent accepted, a blind eye being turned towards the practice. The scarcer a commodity was, the more expensive it became; therefore penicillin scarcity was nurtured by a system which controlled both its price and distribution.[68] During the post-war years, the state avoided a free market of scarce, basic foodstuffs (the rationale behind ration cards) while the black market was regulated by a capitalist economy's relationship between supply and demand: the control of prices—limiting the extent of capitalism—shifted capitalist norms to a black market in which it was fully applied.

The practices that surrounded the paths taken for obtaining penicillin suggest smuggling was an open secret, as publicly recognised as the black market of basic foodstuffs. Networks of support protected and maintained penicillin smuggling, sharing this space for illegal trade with food; space that scarcity had created in the highly controlled nations countries became in the time immediately following war. Its wonders perceived as miraculous, actually obtaining doses of penicillin during the 1940s was a miracle in itself. Poverty and illness, already regarded with apprehension by the Franco dictatorship, joined forces with illegal trade. By the

1950s, however, circumstances had changed. While the dictatorship retained autarchy as an official industrial and economic policy, trade relationships with neighbouring countries developed that changed not only policy options but also those of individuals. A person could be treated with either legal or illegal penicillin, or both. The coexistence of a legal and illegal market of penicillin suggests enduring corruption, which the political decision to create two factories could not eliminate. This duplicity, which has come to be regarded as a fundamental part of Franco's governing style, became a powerful agent in the early biography of penicillin within Spain.[69]

Notes

1. Antonio Cazorla Sánchez, *Las políticas de la victoria: la consolidación del Nuevo Estado franquista, 1938–1953* (Madrid: Marcial Pons Historia, 2000), 189.
2. Expediente de contrabando de mayor cuantía nº 9 (Large Quantity Contraband File No. 9), Delegación de Hacienda en la Provincia de Cádiz, Tribunal Provincial de Contrabando y Defraudación, Archivo de Hacienda, grupo (1)2.10, legajo19426, Archivo General de la Administración Alcalá de Henares, Madrid.
3. A pattern of tobacco smuggling activity in an earlier period in Spain is well documented in Mercedes Cabrera, *Juan March 1880–1962* (Madrid: Marcial Pons, 2011), Chap. 1.
4. On injections see Hobby, *Meeting the Challenge*, 293, no. 68; Bud, *Triumph and Tragedy*, 62–3.
5. Maria Rentetzi, 'Packaging radium, selling science: Boxes, bottles and other mundane things in the world of science', *Annals of Science* 68 (2000): 375–99, and references therein. See also Maria Rentetzi, *Trafficking Materials and Gendered Experimental Practices: Radium Research in Early 20th Century Vienna* (New York: Columbia University Press, 2008).
6. Óscar Rodríguez Barreira, 'Cambalaches: hambre, moralidad popular y mercados negros de guerra y postguerra', *Historia Social* 77 (2013): 149–74.
7. Carlos Barciela is the pioneering and permanent reference: Carlos Barciela, ed., *Autarquía y mercado negro: el fracaso económico del primer franquismo, 1939–1959* (Barcelona: Crítica, 2003). See also Thomas Christiansen, 'Crime and punishment in the black market in Spain, 1940–1950: An analysis of the Fiscalía de Tasas', *Ager* 4 (2005): 63–99. Testimonies and reminiscences are from Beatriz Peña González, *El estraperlo en Cádiz: la realidad social* (Cádiz: Quorum, 2004), 132–86.

8. Miguel Ángel del Arco Blanco, 'El estraperlo: pieza clave en la estabilización del régimen franquista', *Historia del Presente* 15 (2010): 65–78. Miguel Ángel del Arco Blanco, 'Hunger and the consolidation of the Francoist Regime (1939–1951)', *European History Quarterly* 40 (2010): 458–83.
9. Miguel Ángel del Arco Blanco, *Las alas del Ave Fénix: La política agraria del primer franquismo (1936–1959)* (Granada: Comares, 2005).
10. For a memory of the time, see Joan Guillamet, *Tots hem fet estraperlo* (Barcelona: Dima, 1968), 76–81.
11. On the early manufacturing and circulation of nylons see Pap Ndiaye, *Nylon and Bombs: DuPont and the March of Modern America* (Baltimore: Johns Hopkins University Press, 2007).
12. *ABC Sevilla*, March 12, 1944, 10.
13. For an account of smuggling in Spain in recent times, see F. Sánchez Agustí, *Espías, contrabando, maquis y evasión: la Segunda Guerra Mundial en los Pirineos* (Lleida: Milenio, 2010).
14. On the scarcity and early distribution of penicillin in the USA, see David P. Adams, *The Greatest Good to the Greatest Number: Penicillin Rationing on the American Home Front, 1940–1945* (New York: Peter Lang, 1991).
15. *La Vanguardia*, March 12, 1944, 3; *ABC Sevilla*, March 12, 1944, 10; in Cabrera, *Juan March*, Chap. 1, Oran was mentioned as one of the seaports from which tobacco smuggling occurred.
16. *ABC*, March 12, 1944; On Oran and Algiers as settings for tobacco smuggling in Spain, see Cabrera, *Juan March*.
17. Details on the case are in *La Vanguardia*, March 11, 1944, 6; the report stated that the penicillin was 'like that obtained' by this Brazilian institution.
18. Bud, *Triumph and Tragedy*.
19. *ABC*, April 23, 1944.
20. *La Vanguardia*, March 12, 1944, 5; A long, full-page report was published with many details on what penicillin was and how it had been obtained and tested in Britain.
21. See Marlene Burns, 'Scientific research in the Second World War. The case for Bacinol, Dutch penicillin', in *Scientific Research in World War II. What Scientists did in the War*, ed. Ad Maas and Hans Hooijmaijers (London and New York: Routledge, 2009), 44–61.
22. *La Vanguardia*, April 18, 1944, 10.
23. Mariano Jiménez Casado, *Doctor Jiménez Díaz. Vida y obra* (Madrid: Fundación Conchita Rábago de Jiménez Diaz, 1993), 348–55.
24. 'Lógrase encontrar penicilina', wrote Marañón. See Gregorio Marañón, 'Comentarios sobre un caso de meningitis estafilocócica', *Medicamenta* 3(80) (June 1945): 361–2.

25. Federico Tello Valdivieso, 'Aportaciones al conocimiento de las micosis bronco-pulmonares', *Revista Clínica Española* 20 (1946): 1–24; Tello reported on treatment with potasium yodide at his practice in Zaragoza.
26. 'Llega a Madrid el primer envío de penicilina para su distribución en España', *ABC Sevilla*, September 11, 1944. The first mention I have found in newspapers is in *La Vanguardia*, November 12, 1943, 5; this is about penicillin in Buenos Aires, Argentina.
27. Chester S. Keefer, 'The present status of penicillin in the treatment of infections', *Proceedings of the American Philosophical Society* 88 (1944): 174–6.
28. Hobby, *Meeting the Challenge*, 143–5.
29. 'El control de la penicilina', *La Vanguardia*, September 22, 1945, 2.
30. In a 1946 health authority report the 'official currency budget' for importing penicillin was regarded as inadequate, which suggests the state was responsible for payments (see Chap. 3).
31. On this argument, see Christiansen, 'Crime and punishment'.
32. *ABC Sevilla*, November 9, 1944.
33. *ABC*, September 21, 1944, 9.
34. *La Vanguardia*, September 21, 1944, front page; *ABC*, September 22, 1944, 13; on US bureaucracy's attitude to penicillin when it was still scarce, see Adams, *The Greatest Good*.
35. 'Decreto de 5 de junio sobre elaboración e importación de especialidades farmacéuticas', *Boletín Oficial de Estado*, June 26, 1940.
36. 'Llega a Madrid el primer envío de penicilina para su distribución en España', *ABC Sevilla*, September 11, 1944. The Commission could have been created in 1943, according to Guillermo Olagüe de Ros, *Del uso de la retórica en el discurso científico: a propósito de los programas de trabajo de Fidel Fernández Martínez (1890–1942) y Eduardo Ortiz de Landázuri (1910–1985). Discurso leído en el acto de recepción pública* (Granada: Real Academia de Medicina y Cirugía de Granada, 2001), 33.
37. *Boletín Oficial de Estado* 303, 1944, Orden de 4 de noviembre de 1944 por la que se dan normas para el uso de las 'penicilinas'.
38. *ABC*, December 17, 1944.
39. Del Cura, María Isabel and Rafael Huertas, *Alimentación y enfermedad en tiempos de hambre: España, 1937–1947* (Madrid: CSIC, 2007).
40. Christiansen, 'Crime and punishment'.
41. Christiansen, 'Crime and punishment'.
42. The owner of the public house was a patient of Jiménez Diaz, according to Jiménez Casado, *Doctor Jiménez Díaz*, 350; see also the personal recollections of a rural MD, J. Chamorro, *Recuerdos de una vida* (Oñate: Diletante, 2002), 28, 46–8.
43. Bud, *Triumph and Tragedy*, 59.

44. On the new styles and policies of prescribing in the USA, see Jeremy Green and Elisabeth S. Watkins, eds., *Prescribed: Writing, Filling, Using and Abusing the Prescription in Modern America* (Baltimore: Johns Hopkins University Press, 2012); on scientific marketing, see Jean-Paul Gaudillière and Ulrike Thoms, eds., *Pharmaceutical Firms and the Construction of Drug Markets: From Branding to Scientific Marketing*, special issue of *History and Technology* 29(2) (2013). https://doi.org/10.1080/07341512.2013.828867.
45. 'Moción de 11 de septiembre de 1946', Ministerio de la Gobernación, Dirección General de Sanidad, copia con registro de entrada en el Instituto Nacional de Industria de 8 de enero de 1947, Archivo Histórico del Instituto Nacional de Industria (Madrid), legajo 270, documento 7 (1946–1947); the Instituto Nacional de Industria (INI, the National Institute of Industry) was created in 1941 to promote state participation as an entrepreneur and state investments in the Spanish industrialisation process after the Spanish Civil War. During the first decade of Franco's regime, the INI was the main authority in industrial development, creating many industries and promoting state monopolies in several sectors. On the history of the INI, see P. Martín-aceña and F. Comín, *INI. 50 años de industrialización en España* (Madrid: Espasa Calpe, 1991).
46. CEPA, advert: 'Medio millón de frascos de penicilina G sódica...fabricada en Estados Unidos', *ABC*, December 31, 1950, 21.
47. On the film, see J. G. Sánchez and E. G. Sánchez, 'Antibióticos y cine: El tercer hombre y Mercado prohibido', *Revista Española de Quimioterapia* 17 (2004): 223–5.
48. 'Ley de 20 de diciembre de 1952 sobre Contrabando y defraudación', *Boletín Oficial del Estado*, December 24, 1952.
49. CEPA: *Memoria del ejercicio 1953* and *Memoria del ejercicio 1954*. Archivo del Servicio de Estudios del Banco Urquijo. Biblioteca de Ciencias Sociales y Jurídicas, Instituto Juan March-Universidad Carlos III de Madrid de Ciencias Sociales, Getafe, Madrid.
50. On the black market of wheat during the same period, see Carlos Barciela, 'Crecimiento y cambio en la agricultura española desde la Guerra Civil', in *La economía española en el siglo XX. Una perspectiva histórica*, eds. J. Nadal, A. Carreras and J. Sudriá (Barcelona: Ariel, 1987), 258–79. For the illegal trade of other goods in the same period, see Sánchez-Agustí, *Espías, contrabando*.
51. E. Medina García, 'Contrabando en la frontera de Portugal: orígenes, estructuras, conflicto y cambio social' (PhD diss., Madrid: Universidad Complutense de Madrid, 2004).
52. Delegación de Hacienda de Santa Cruz de Tenerife, Junta Administrativa de Contrabando y Defraudación, Archivo de Hacienda, grupo (1)2.10, legajo19426, Archivo General de la Administración Alcalá de Henares, Madrid.

53. Fondos documentales del Tribunal Económico Administrativo Central, libro de minutas de fallos de expedientes de contrabando 1955, n. 168. Ministerio de Hacienda, Biblioteca central, Madrid, hereafter Archivo del Ministerio de Hacienda.
54. Archivo Ministerio de Hacienda, libro de munutas y fallos 1955, n. 212.
55. Archivo Ministerio de Hacienda, libro de munutas y fallos 1956, n. 410 (110103).
56. Archivo Ministerio de Hacienda, libro de munutas y fallos 1956, n. 424.
57. Archivo Ministerio de Hacienda, libro de munutas y fallos 1956, n. 44.
58. Archivo Ministerio de Hacienda, libro de munutas y fallos 1955, n. 150.
59. Archivo Ministerio de Hacienda, libro de munutas y fallos 1955, n. 189.
60. Archivo Ministerio de Hacienda, libro de munutas y fallos 1954, n. 345.
61. Archivo Ministerio de Hacienda, libro de munutas y fallos 1955, n. 415.
62. Delegación de Hacienda de Barcelona, Tribunal provincial de Contrabando y Defraudación, Archivo de Hacienda, grupo (1)2.10, legajo19426, Archivo General de la Administración Alcalá de Henares, Madrid.
63. Delegación de Hacienda en la Provincia de Granada, Tribunal Provincial de Contrabando y Defraudación, Archivo de Hacienda, grupo (1)2.10, legajo19426, Archivo General de la Administración Alcalá de Henares, Madrid.
64. Tribunal provincial de contrabando y defraudación de Málaga, Exp 738/54 (1)2.10 Legajo 19428. AGA, Alcalá de Henares, Madrid.
65. Antonio Cazorla Sánchez, *Fear and Progress: Ordinary Lives in Franco's Spain, 1939–1975* (Oxford: Wiley, 2009). On the end of the 1950s onwards, see Nigel Townson, ed., *Spain Transformed: The Franco Dictatorship, 1959–1975* (Basingstoke: Palgrave Macmillan, 2010).
66. On these ideas of concealment, although in a later period, see Amando de Miguel, *España oculta: la economía sumergida* (Madrid: Espasa Calpe, 1988).
67. On smuggling as a secret trade, see E. Tagliacozzo, *Secret Trades, Porous Borders: Smuggling and States Along a Southeast Asian Frontier, 1865–1915* (New Haven, CT: Yale University Press, 2005).
68. Although Spanish authorities during the dictatorship displayed interest in controlling the black market, the Ministry of Agriculture estimated that illegal trading of wheat accounted for over fifty per cent of the total wheat traded during the 1940s. In the agrarian sector at least, the black economy was hugely significant during that decade. See Carlos Barciela, *Intervencionaimos y crecimiento agrario en España*, in *La nueva historia económica de España*, eds. P. Martín Aceña and L. Prados de la Escosura (Madrid: Tecnos, 1985), 285–316. Also Christiansen, 'Crime and punishment'.
69. Something similar is discussed by Jordi Catalán, *La economía española y la segunda guerra mundial* (Barcelona: Ariel, 1995).

CHAPTER 5

Modern Times: Screening Antibiotics and the Factory Line

This chapter is an account of the screening programme in search of new antibiotics established by CEPA (Compañia Española de Penicilinas y Antibióticos) and Merck in Madrid in 1954. An exploration of the genealogy for such a programme, its narratives and practices, shows that the main inspirations were the factory system of production, and Selman Waksman's research agenda on microorganisms of the soil. The relationship between industrial production of antibiotics and the research programme intended to identify new candidate drugs is examined, and I suggest there were organisational similarities. Moreover, the research objects and tools originated—both materially and conceptually—from industrial production: a line of artisanship constructed to obtain a product through the combined effort of the production line workers. Following the style developed by Selman Waksman at Rutgers University, the screening programme evaluated samples manually, and microbiological skills were enhanced with every test. The Madrid team's practice of using instructions from Merck led to the circulation of knowledge and practices, including research material and microbiological methods.

At 6:30 a.m. on December 23, 1954, Sagrario Mochales, a young woman with a university degree in natural sciences, took the underground to Atocha station in Madrid. There she boarded a truck—similar to the military ones covered in canvas used to transport soldiers—to go to the CEPA factory for her first day of work. Every day, CEPA workers took this transportation, provided by the factory, to travel from Atocha

M. J. Santesmases, *The Circulation of Penicillin in Spain*,
Medicine and Biomedical Sciences in Modern History,
https://doi.org/10.1007/978-3-319-69718-5_5

square to the facilities at Méndez Alvaro, an industrial neighbourhood in southeast Madrid. After boarding with the help of a small ladder they remained standing for the short daily journey.

She had been selected by the director of CEPA's department of quality control to work in a newly created research department: Mochales would be screening for new antibiotics alongside a physician and a chemist, both male. The three of them were beginning their training as interns in the CEPA laboratories immediately before Christmas 1954. Mochales and the technical director of the factory, Carmela Viejo, were the only women with university degrees working at the factory until 1966, when a young female graduate joined the CEPA research programme.[1]

Drugs manufacturing belonged and still belongs to an idea of invention and production closely related to that depicted in the film *Modern Times* (Charles Chaplin, 1936). Industrial production was done on the assembly line, a set of methods and tools generating products, the healing capacities of which in the treatment of infections also generated health standards, ways of inventing and styles of looking at living material. The production of drugs led to a systematic way of searching for them. This systematisation included methods and tools that circulated from one place to another.

This was the case with the programme to search for new antibiotics set up in Madrid in the early 1950s. During that decade Spain began to recover from the devastating effects of the Civil War and the early decade of Franco's dictatorship, a time of hunger and poverty combined with corruption and a black market of basic products—bread and drugs among them—and national isolation.[2] Norms and policies of the Franco dictatorship combined with foreign industrial relationships to provide the means and grounds for experimental research.[3] In the case of the programme explored here, CEPA's interest in conducting research at the Spanish penicillin firm appears to have coincided with the interest of Merck in diversifying its research activities, while also keeping the line of research for new antibiotics started by Selman Waksman active.

This chapter explores the origins of this collaborative research programme established by Merck and CEPA in 1954. The agreement to establish a screening programme was the third in a series: the first led to the industrial manufacturing of penicillin; the second, of streptomycin. From 1954 on, this screening programme, established in CEPA's bottling plant in Madrid and fitted with an industrial style of work, created a research agenda based on utilising microbiological skills to detect antimicrobial activity in soil samples.[4] Such research performance became part of the international environment being created as part of recovery policies in Europe during the late 1940s, to promote productivity, both in industry and scientific research, after the Second World War. As a programme, the CEPA–Merck research agreement and the research agenda it established exhibited a strong influence from the factory system of production, in the organisation and management of a research laboratory situated in a penicillin factory. The participation of woman researchers will also be emphasised and explored, so as to genderise the history of antibiotic research and the practices of microbiology that participated in it, in an environment in which English language was the instrumental mediator.

Post-War Promises

The vanquishing of infections through the industrial manufacture of antibiotics was one of the most eagerly anticipated possibilities of the therapeutic revolution. Following the Second World War, cures for conditions ranging from syphilis to tonsillitis arose as both material therapy, and a metaphor for the healing of such devastation. At a time when penicillin was distributed in Europe and North Africa through diplomatic channels, and new antimicrobial drugs began to be industrially manufactured in various European countries, the research laboratory remained the main space of practices in the trajectory of penicillin.[5] After the marvel and, at times, unfulfilled promise of sulfa-drugs, penicillin, in all is marketed salt forms, became the first 'antibiotic'.[6]

Research was deemed an essential foundation for European economic and social recovery, therefore the promotion of research was a core principle for the new international scientific and technological organisations engaged in European reconstruction. The United Nations Educational, Scientific and Cultural Organization (UNESCO) was created in 1945 and the Organisation for European Economic Cooperation (OEEC) in 1948. The promotion of scientific and technical education and research was pursued by both international agencies, and their reports would soon become influential agents in the design of science and technology policies. The long post-Second World War period became—in addition to the era of the promise of an infection-free world and closely related to this prospective aim—the era of scientific and technological research. As the early office for distributing the Marshall Plan budget in

Allied countries, OEEC (later renamed the Organisation for Economic Co-operation and Development [OECD]) was influential in creating research policy as it is understood today: for the funding of scientific and technical training and research as instrumental paths to industrial development on a national basis.[7]

The recovery agenda therefore played a part as an agent creating expectations, inventing a future of promises. Policies toward 'development'—a term of the post-war era—generated futures of technology associated with industry and productivity, toward improving ways of creating goods; goods often used to improve health and, in this case, to cure. These research and development policies became a means of increasing economic, industrial productivity. European science and technology policy was originally conceptualised and designed at the European Productivity Agency of the OEEC. Technology as ideology significantly influenced penicillin distribution policy, as Slovo Łotysz has recounted in the case of Poland, while the promotion of research as a path to development played a role in establishing the screening programme in Madrid.[8] Geographies of knowledge became entangled, as Gabrielle Hecht has phrased it; the geopolitics of such entanglement including the creation in 1948 of the World Health Organisation (WHO), with its authoritative agenda to promote health policies and drug development.[9]

This geopolitical and epistemic interaction between post-Second World War policies and the trajectory of antibiotics was based in the research laboratory. In both the search for new antibiotics and determining their mode of action, laboratories became the place where recovery was made possible through new drugs.[10]

An exploration of the genealogy of the CEPA–Merck screening programme, its narratives and practices, shows that the main inspiration for this programme was not only the mentioned factory system of production but also Selman Waksman's research agenda on microorganisms of the soil. I suggest the screening programme in search of new antibiotics was organised along industrial lines. Research was motivated and organised according to industrial management, in what could be described as a counter-linear model. That is, it shows the origins of research in the factory rather than the opposite way round: the building of the factory after some research had been carried out.[11] As Robert Bud has argued, after the Second World War research carried out by industrial firms developed particular organisational methods and orientations that would have influenced large projects being carried out by academic and medical scientists.[12] I would like to further this argument and suggest the influence of industrial production itself in the organisation and management of, in this case, research on microorganisms.

The research programme analysed here included training, personal instruction and repeated manipulations while keeping laboratory skills at the core of workplace practices. These were as visible as the factory in which they were applied, developing a research project and assembling a production line of samples from which antimicrobial activity would hopefully be detected. Research objects and tools of this screening programme in search of antibiotics came, both materially and conceptually, from industrial manufacturing.[13] And it was in the factory where the research programme and the materials and research participants occupied physical space, part of the factory itself and following industrial patterns of action. So it was not only that the devices used had been manufactured by an industrial firm and became part of the work at the bench, design of the research programme itself was a product of an industrial project, that of antibiotic production in Spain.

The practice of sample screening had been set up in Selman Waksman's laboratory at Rutgers University. The routine of repeated manipulation of soil samples in search of new *Actinomycetes* was far from being the 'uncreative testing techniques' that William Kingston has described.[14] Samples were evaluated manually, and microbiological skills enhanced and expanded with every test. By applying instructions, following recipes elaborated elsewhere, the Madrid team's work was a product of the circulation of knowledge and practices. It demonstrated training and knowledge production in microbiology and an industrial manufacturing style, by making research materials and methods of microbiology travel between different geographical and professional settings.

The Factory and the Research Line

This research partnership between CEPA and Merck was set up immediately after the penicillin factory in Madrid was constructed. The research programme originated from a factory—at Merck headquarters in Rahway, New Jersey—then developed in a laboratory and moved to the factory in another country. This move reinforced the commercial agreement between CEPA and Merck while constructing a research culture in an industrial setting. Research was part of a process of industrialisation, which at that particular time in Madrid meant the construction of factories to make products in Spain with manufacturing processes invented abroad. That is, manufacturing procedures for both research and antimicrobial drugs were brought from a US firm by a Spanish one.[15]

Post-war ideology regarding the promise of scientific research included promoting—designing and funding—technological developments that would eventually defy resistance to the action of penicillin and other antibiotics marketed through the late 1940s and early 1950s. Resistant bacteria took part in the search for new antibiotics, not only as problems to be addressed but also as biological responses to antibiotics.[16] The agendas of both microbiological research and the pharmaceutical industry included the locating of new therapeutic agents to combat infections.

With this screening partnership, Merck acknowledged the 'serious need to meet the challenge of antibiotic-resistant pathogens'. In the words of the Merck Research Laboratories directors, Harold Boyd Woodruff and Lloyd E. MacDaniel, 'antibiotics, which are more efficient as antibactericidal agents, which can penetrate to calcified lesions or the deepest recess of the heart, are needed for chronic diseases such as tuberculosis or bacterial endocarditis'.[17] Although this need for new antimicrobials was primarily about resistance, there still existed infections for which, according to the narrative of Merck's microbiologists, efficient drugs had not yet been discovered. Resistance became a powerful challenge to be faced by manufacturing new drugs, an industrial feedback loop that took full advantage of the circumstances promoting further research and increasing profits. Woodruff and MacDaniel went on to describe the underpinning logic of 'conventional microbiological techniques' used in the search for *Actinomycetes*, and the soil samples put through what they termed the 'random screening method'. New antibiotics were found 'every year', and would 'displace existing antibiotics as physicians study them as therapeutic agents'. Not only resistance but novelty itself appears to have held merit in the finding and manufacturing of antibiotics.[18]

As the directors Woodruff and MacDaniel—both expert microbiologists—phrased it at a 1957 symposium, it was the antibiotics 'discovered by random methods, which carry today the burden of chemotherapy'. By 1956 'the value of antibiotic manufacture' in the USA was 242 million dollars and 1.5 million pounds in weight. In spite of past accomplishments, antibiotic-resistant pathogens were a 'serious' challenge.

'Without question, there is a need for new clinically useful antibiotics', as was shown by the existence of screening programmes in 'practically every large pharmaceutical company and many university laboratories'.[19]

The scope of antimicrobial therapy appeared limitless following the Second World War. The already known and manufactured goods—penicillin and, slightly later, streptomycin—generated further promises, as there were still infections neither could cure. These were the two sides of antibiotic resistances: they cautioned against the overuse of antibiotics, but at the same time created the research agenda. New antibiotics remained worthy objects to hunt for.

It was in this environment of challenges and needs that the CEPA–Merck research laboratory was established. This enabled Merck to spend the royalties provided by the sale of penicillin and streptomycin in Spain. CEPA paid these royalties to Merck in Spanish currency, and Merck used them to establish and develop the research programme, thereby relocating the early steps of its screening from Rahway to Madrid.[20] The methods and procedures of the programme made the journey from the USA to Spain at a time when they had already proved their usefulness: streptomycin had been found, tested and manufactured through these methods. As head of the early screening programme at Merck, in collaboration with Rutgers University, Waksman had been awarded the Nobel Prize for Medicine in 1952. The CEPA research laboratory group trained in the screening of new antibiotics, experienced an expectation of participation in new discoveries, and the scientists involved trusted the screening method as a reliable scientific technique.

The Practices of Screening in Microbiology

Previous pharmaceuticals produced in Spain had been vaccines and sera, both made during the first third of the twentieth century at the Instituto de Higiene, created and funded by the state, and at private firms.[21] The chemical industry was dependent on supplies bought abroad, and dyestuffs and fertilisers were also imported for the textile industry.[22] Thus, the career of Paul Ehrlich in the German pharmaceutical industry, which led to the manufacture and use of the arsenical Salvarsan and later the less toxic neosalvarsan, and that of Gerhard Domagk in the development of sulfa-drugs, pertained to developments within the German pharmaceutical industry that originated from the chemical expertise.[23] The case presented here shows a path from industry to research; research was a result of industrial production, a product of industrialisation. The research area was microbiology, that is, expertise in handling and isolating microbes and their prospective antagonists.

Training in microbiology and culturing techniques took place during the process of creating and setting up the screening programme.[24] Screening for new antibiotics was in tune with the predominant rhetoric in the West from post-Second World War and throughout the Cold War, which proclaimed research as the path toward economic recovery and industrial development.[25]

This post-war period was also the era of screening, of systematically testing cultures, samples, techniques as well as patients, in order to discover improved methods of diagnosis and treatment. Industrial screening for therapeutics and the clinical screening of patients for a disease or a disorder shared a number of features, such as a systematic search for something otherwise undetected, a kind of blind search. Screening for latent disorders in living beings, such as PKU in new-borns and cervical cancer, required additional sensitivities, both materially and symbolically.[26] All screening, however, displayed an interaction between industrialisation, as a way of distributing work along an assembly line, and that of screening as an assembly line of testing, of systematically looking for, in the case of antibiotics, antimicrobial activity in samples from which a new drug could be detected and isolated. It appears this system of screening had been modelled on industrial organisation. Not only was there a sequence of tasks, but also an integrated set of procedures for the mass production of new samples of antibiotics. Both the means—a production line—and aims—mass production—of the antibiotic screening programme appeared as inspired by the factory system.[27] An interdisciplinary team, composed of graduates in medicine, chemistry and natural sciences, took part in the search for new antibiotics, each assigned to particular tasks in the research line.

Antibiotic screening was understood as a linear way of conducting tests, collecting evidence and obtaining proof of either the presence or absence of activity. In this sense, John Lesch's book, *The First Miracle Drug*, is suggestive, and convincing, regarding the instrumental role played by research in the development of sulfa-drugs, which created a method composed of systematically testing reagents resulting from the chemical modification of a given molecule. Although not described as screening, it was a system of production of prospective new anti-infective chemical products. Screening as such was based on, or inspired by, this industrial organisation of tasks.

Microbe cultures, however, were very different. Although they did not require the technical expertise that the organic and biological chemistry of dyestuffs and sulfa derivatives did, their handling required particular skills and sensitivities.[28] Microbiological research involved the observation of living material by an expert, trained eye. Rather than structure, or any other chemical characterisation such as composition and prospective synthesis, it was biological activity that required identification. Microbiological practices of screening entailed artisanal work: careful examination of microbe cultures, at the rings around them, their shape and intensity. Chemical identification, on the other hand, was post-industrial technology mediated by a diversity of devices. Screening as practice in the programme analysed in this chapter, was a set of techniques to search for the enemies of microbes.

The tasks at the CEPA laboratory were basically microbiological, including the collection of soil samples and the isolation of microbes whose antibiotic capacity would be tested against a list of known bacteria. The preparation of concentrated samples of an antimicrobial and its toxicity characterisation would be carried out. Further chemical identification of the active product in the extract would be completed at the Merck Research Laboratories in Rahway.

The identification of a sample's antibiotic capacity and characterisation of microbes of the species *Actinomycetes* required the careful handling of Petri dishes and examination of the ring surrounding an active sample. Merck, meanwhile, concentrated on the technical skills of chemical isolation and identification of composition and structure, for which the synthesis of the molecule under study was often required. Although microbiology came to replace chemistry as the discipline involved in finding new antibiotics from the discovery of penicillin onwards, chemistry remained a participant in drugs research and production during the antibiotic era.

Research at Merck, Selman Waksman and Streptomycin

It was in the early 1930s when Merck began to create a chemical research laboratory, and hired Raldolph T. Major, an organic chemist graduate from Princeton, as director. The firm approached pharmacologist Alfred Newton Richards, from the University of Pennsylvania School of Medicine, for advice on establishing a pharmacological laboratory. John Swann has argued that Richards's role—long before his 1941 appointment as chairman of the Committee of Medical Research by President Roosevelt, to advise the Office of Scientific Research and Development as part of the US war effort—was to be a liaison between Merck and the academic community. Receiving a recommendation from a colleague, Richards suggested Merck hire Hans Molitor, assistant professor at the Pharmacological Institute of the University of Vienna. Richards conducted negotiations to this effect, and Molitor arrived in the USA in the summer of 1932. Under Molitor's leadership the newly created Merck Institute for Therapeutic Research in Rahway became a 'leading industrial research centre' in the USA.[29]

The Merck Institute carried out research on a wide variety of therapeutic agents, including hormones, anaesthetics and vitamins. In 1937, the Institute entered the field of chemotherapy, manufacturing and testing sulphanilamide derivatives and slightly later sulphapyridine. This new sulpha derivative was distributed to US physicians in March 1939 for the treatment of around 15,000 patients, either in clinical trials or emergency cases 'during the pneumonia season'. The research capacity and structure of the Institute expanded, with a new building for the investigation of synthetic compounds completed in 1940, a new building for chemotherapy in 1949, a poultry nutrition building in 1950 and a new laboratory administration building in 1952. The impressiveness of this 'research complex' was celebrated in 1953, with an event attended by numerous government officials, business leaders and scientists, including Alexander Fleming and Selman Waksman.[30]

Merck soon expanded into microbiology. The methods they employed for detecting new antibiotics appear to have been based on the action of those previously detected and isolated. Soil was known to be full of antagonistic microorganisms, an arena of conflict: life against life, as Paul Vuillenmin had phrased it in 1889.[31] In the research laboratory the Petri dish reproduced this microbiological fighting on a small, observable scale.[32] In 1938, Merck hired Selman Waksman as a consultant and established a fellowship in fermentation studies at Waksman's laboratory.[33]

Soil microbiology had been Waksman's early research project as a graduate student, the subject through which many of his students were introduced to research, and the platform from which systematic research on the antagonistic action of soil microbes took place.

These biological products were everywhere in Waksman's research landscape. *Actinomycetes* were the microorganism that Waksman was best acquainted with. By the 1910s, Waksman and R. E. Curtis had obtained 'hundreds' of *Actinomycetes* from soil. They were 'fascinated' by their structure and function and, once they had been characterised, established the bases for regarding such microorganisms as ones that limited bacterial growth in soil through what they described as an 'antagonistic nature', the capacity of one microorganism to kill others, an idea at the very heart of bacteriology.[34] By then, Waksman was lecturing on soil bacteriology at Rutgers Agricultural College, and his research agenda revolved around what could be called soil ecology: the study of the relationship between different types of microorganisms in the soil.[35]

Waksman's early contributions to Merck were on the submerged fermentation of lactic and citric acids. It was shortly after Waksman's former student, René Dubos, detected an antimicrobial microorganism that Merck established a fellowship in Waksman's laboratory for research on antibiotics. Merck signed an agreement whereby the firm would provide chemical and mass production support for any new, promising antibiotic detected by Rutgers' microbiologists.[36] As Waksman and researchers in his group identified and conducted preliminary tests on a new antimicrobial substance, Merck chemists purified it, and determined the chemical structure. Merck pharmacologists tested its toxicity and therapeutic activity in animals, and bacteriologists and engineers developed the project, establishing methods for large-scale production.

Together with Woodruff, Waksman isolated an antibiotic from *Actinomycetes* which proved highly toxic to laboratory animals. After this 'failure', a 'comprehensive programme of screening *Actinomycetes*' was launched, as they had noted a diversity of antimicrobial activity, chemical features and toxicity. With hopes of filling the Gram-negative bacteria therapeutic 'gap' left by penicillin, the objective was to secure the relevant antagonists.[37]

The relationship between Merck and Selman Waksman's laboratory developed over the years. A number of Waksman's former students were hired in Merck's research laboratories, where facilities were available for any development the Rutgers group might achieve.[38] Among the earliest patents were those secured by Waksman and Albert Schatz for streptomycin in 1943, which proved to be active against 'certain forms' of tuberculosis and *Hemophilus*.[39]

By that time, on the basis of an agreement with Rutgers, Merck held patent rights for the production of all substances identified and isolated in the framework of this agreement, and for others when research funds from Merck were granted to Waksman's laboratory to search for new antibiotics. Waksman later renegotiated the agreement: Rutgers obtained patent rights on both streptomycin and actinomycin.

The agreement with Merck was revised so other companies could produce streptomycin. Merck accepted a non-exclusive license and in 1946 assigned its patents associated with streptomycin to the Rutgers Research and Endowment Foundation; two years later eight US companies were manufacturing it.[40] What had been established in 1938 as an apparently modest consultancy contract on citric acid, yielded 12 million dollars through antibiotics over the next forty years.[41] Waksman would later argue that detection of streptomycin would not have been the huge success it was without the help of Merck, 'an industrial organisation that took over a major part of the pharmacological evaluation' of the new drug.

Industrial Connections

After developing the efficient production of penicillin as part of the war effort, Merck now owned patent rights to penicillin and streptomycin manufacture and had started to license them outside the USA. This was the stage for Merck's expansion to Spain, the commercial agreement signed with a group of Spanish chemical and pharmaceutical companies belonging to Banco Urquijo in 1948, and the building of the penicillin production plant in Madrid, as detailed in Chap. 3.

At that time, the 'comprehensive programme of screening *Actynomicetes* for their ability to produce different antibiotics' had been successfully developed by Waksman's research group at Rutgers.[42] The group collected soil samples, plated them out, isolated *Actinomycetes* and repeatedly tested for antibiotic activity against non-pathogenic strains of mycobacteria, in the hope of finding new substances active against pathogenic strains.[43] This seemingly simple project, the description of which could evoke routine and repetitive activity, was dutifully maintained at Waksman's laboratories even through the 1950s, when there seemed to be a consensus—apparently not shared by Merck microbiologists—that no new antibiotics were required, given the amount already being manufactured following successful clinical trials.[44] However, the increasing resistance of bacteria to the activity of many antibiotics spurred on the hunt. The systematic study of soil preparations as a repetitive, planned programme was part of a microbiological strategy, closely associated with the meaning of 'antibiotic', a term coined by Waksman himself.[45]

Previous commercial agreements between Merck and CEPA were managed by CEPA's director, Antonio Gallego. Once the penicillin factory in Madrid was in operation, Gallego participated in establishing this 'new research endeavour' at CEPA's laboratories using income from penicillin and streptomycin manufacturing licences.[46] The documents relating to this agreement, as well as testimonies provided by former CEPA colleagues, suggest Gallego was fully aware of the research programme developed by Merck's Research Laboratory and Institute for Therapeutic Research, set up in 1938.[47] By the time the agreement was signed, in September 1954, a streptomycin plant was in full production; Waksman had attended the inauguration in Madrid on the tenth anniversary of the drug's 'discovery' (Fig. 5.1).[48]

Tasks for 'Screening'

The contract signed in 1954 stated that CEPA intended to establish a screening programme for the discovery and development of antibiotics.[49] What screening actually meant at that point was not clear to the young graduate in natural sciences, Sagrario Mochales, who applied for a post in the CEPA laboratory.[50] For Merck and CEPA and this particular project it described the systematic testing of soil samples in search of *Actinomycetes* possessing antibiotic activity against Gram-negative bacteria. At that time Merck was interested in broadening the search for new antibiotics 'but the cost of doing so and the interference with ongoing research projects, were inhibitory'.[51] An opportunity arose when Antonio Gallego suggested a collaborative research project. H. Boyd Woodruff, head of the research section and later director of the department of microbiology and natural products isolation at Merck, had previously trained with Selman Waksman at Rutgers, publishing jointly authored papers on microorganisms isolated from the soil during the 1940s. For Merck this was an opportunity to initiate new research at Rahway, including the development of drugs to destroy microbial dental plaque and exploring the antigenic properties of bacterial polysaccharides.[52]

The purpose of this cooperative project was to study new soil samples, collected from as many places as possible 'by Merck sales agents' and sent to Madrid for testing. Antonio Gallego was involved from very early on, and formalised the agreement with Randolph T. Major from

Fig. 5.1 Selman Waksman with Antonio Gallego, left, and José Luis Gallego, right, at the CEPA streptomycin factory, Aranjuez 1954. Courtesy of Roberto Gallego

Merck Research Laboratories. Lloyd E. McDaniel from Merck and Justo M. Mata from CEPA were appointed directors of the programme.[53]

In order to search for new antibiotics in new soils, the Spanish group had to learn culturing techniques, and methods of treating and testing samples. CEPA medical researcher Sebastian Hernández travelled to Rahway for training in the programme's laboratory practices. Equipment, reagents and methods for isolating and testing soil extracts were 'transferred' from Rahway to Madrid, and McDaniel and Woodruff often made the same journey, particularly as the laboratory and screening line were being assembled. This transfer of skills, equipment and knowledge continued to ensure the programme was established to Merck standards.[54]

By 1956, a year after the research team had been assembled, activity at the CEPA research laboratory included the isolation of *Actinomycetes*, the testing of isolates against a set of bacterial products, the development of active cultures in shake flasks, resistance test on broths and determination of antibacterial spectrums, further production in five-litre flasks and confirmation of previous identification (if any), preparation of concentrates and toxicity tests, and 'determination of protective levels and calculations of therapeutic indices by tests on infected mice'. The agreement included details of the facilities needed to carry out the programme: one laboratory for isolation, one for identification, a chemical laboratory and a biological laboratory. These 'four screening laboratories' needed one supervisor, three to four assistants and a dozen technicians. The agreement even laid down the weekly output requirements of the programme: 12 soil samples, 500 tested isolates, 30 cultures in shake flasks, 6 cultures in fermenters and two new products in animal testing.[55]

This style of producing new substances evokes that of early industrial production, with the integration of workshops under the same roof, and a sole supervisor in charge of the entire process. One person took the sample and sent it to the programme site; a second carried out the first step, from which followed the successive steps that, eventually, produced something new, active and non-toxic. It was a process of progressive selection with a pyramidal structure, the final successful product at the apex, demonstrating the importance of the entire sensitive procedure.

The way in which this research was systematically carried out, as a series of tasks, brings to mind work in industrial, assembly-line manufacturing. But this was achieved through successive artisanal acts: no automatic procedures were used, simply a line of manual tasks. These interactions reveal a landscape in which research agendas and the demands of industrial manufacturing shared both experts and expertise and, perhaps even more relevantly, these two professional spaces shared interests and modes of production.

The project was very much in accord with other screening programmes of the time, such as PKU (phenylketonuria) screening in newborns and screening for cancer viruses and therapeutics, started around the same time. Screening evokes, on one hand, war-time efforts of systematically searching for solutions to military problems or concerns and, on the other, a legacy of chemistry and earlier strategies developed in Germany, which eventually resulted in sulfa-drugs.[56] The screening programme belonged to a genealogy of research agendas based on repeating series of tasks in search of a wide set of biomedical objects.

At the same time, this was a production line which involved a high degree of training. Tasks functioned as a means of production while keeping the activity part of a predesigned line. The production line was created as part of a factory, attuned to the factory system, and the two exchanged means of production. The work distribution of the research programme and factory were mutually consistent: they shared an industrial production culture as well as space in the manufacturing plant.

The soils were a source of microbes as well as their antagonists. This particular practice was a method of searching for new antibiotics, one that specifically integrated the expertise Waksman provided for Merck. CEPA's research laboratory had an artisanal way of working, requiring careful observation of broths for antimicrobial activity and managing the uniqueness of each new sample and microorganism located.[57] The research space was set within the quality control laboratory, the result of interaction between different professional cultures.

At Work

By 1957, when the CEPA research laboratory was working to full capacity, over 95,000 plates of soil samples had been examined; isolation and fermentation tests were conducted on 22,000. In January and June of that year, these samples were contaminated with *Neurospora*. To check their antimicrobial activity, the isolated broths were tested against a select list of bacteria: those resistant to previously developed antibiotics, mostly Gram-negative. Having discarded inactive samples and those active against Gram-positive bacteria, 30% of the samples remained. Further identification by chromatography led to 126 being discarded, and the remaining 41 were tested in a third fermentation against E. coli strains. From these, three new neomycine antibiotics were identified, one xanthomycine and two unknown. Concentrated samples were sent to the research laboratories in Rahway to be purified and identified, as antimicrobials isolated at Rutgers had been as part of the agreement with Merck. The small group of researchers in Madrid, who apparently enjoyed their work, introduced modifications to improve methods of isolation and identification.[58]

The CEPA research laboratory shared its director with the department of quality control: Justo Martínez Mata MD remained in charge for over two decades. Following Sebastián Hernández's return from Rahway—he had originally worked in quality control, and was now head of the research laboratories—three interns were selected: a female natural scientist, Sagrario Mochales; a chemist, Tomás Cubillo; and a physician, Ángel Moreno. As university graduates, they remained in a category apart from the factory workers. Some of the factory 'girls'—as they were called—did move to the new laboratories as technical assistants: 'they were smart girls, already trained in factory work', Sagrario Mochales (2012) remembers. Mochales started as an intern at the research department in 1954 and soon gained a young female assistant. The number of assistants the group employed increased over the next few years as the work became more intensive. At the time of Mochales's selection, a discussion had taken place among CEPA directors about whether a woman should be part of the research group, but Antonio Gallego had finally acquiesced.

Managerial capitalism was compatible with the family firm that to a certain extent CEPA was at that time. Personal relationships were instrumental and remained a source of interns: Mochales was the niece of a clinical expert in infectious diseases, Fernando Baquero, who had a professional and personal relationship with CEPA. She grew up in Madrid, attended one of the very few private schools that did not belong to a religious order, and entered the University of Madrid in the late 1940s. She belonged to a well-educated middle-class family. Her father, who worked for Banco Urquijo, was able to keep his job after the war.

According to Mochales, the screening programme at CEPA was an inexpensive option for Merck: 'We imported instruments and exported results' (Mochales 2012). Mochales believed Spanish researchers were cheaper than US ones, which suggests she regarded US Merck training skills and technical equipment for research as superior to her own capabilities as a scientist and researcher. When reconstructing her work at the research group, Mochales perceives her identity as scientific, even though she was hired as a university graduate 'technician'. She was very young at the beginning, and single; a requirement at the time for any woman to be hired.

The spaces she found to construct her professional identity were the group meetings she joined in Madrid and in the early years at Rahway, during annual visits with Hernández to report results and receive training in new methods and objectives. The programme was expected to be highly productive; for the first decade, however, despite intensive work, not a single new antibiotic was discovered.

New methods would eventually be developed that, with the chemical assistance of Merck, contributed to the isolation and identification of new antibiotics in Madrid.

Circulating Methods: Geographies, the Laboratory and the Manufacturing Plant

By the 1950s, screening as a research strategy, as a programme based on randomness, was being developed by other pharmaceutical industries to continue producing novel and prospective new antibiotics for a market that generated both expectations—the best antimicrobial was yet to come—and anxieties: resistances thought to be created by the antimicrobials themselves. The entire trajectory of screening as a contemporary method was shared by other disciplinary approaches: the best known and one of the most influential was the previously mentioned detection of phenylketonuria (PKU) through the screening of newborns.[59]

Research methods circulated from one pharmaceutical company to another, from one country to another, from a microbiological laboratory to a quality-control laboratory, and between different professional groups. Tools and experimental procedures, emerging in a collaborative research programme, included technical and research skills and entrepreneurial relationships that became stabilised. By providing results—concentrates of prospective new antibiotics—the antibiotic screening programme generated feedback, that is to say, more circulation. On these journeys, crossing national, professional and local boundaries, drugs research led to collaboration and the development of new skills.

The CEPA–Merck screening programme was a research project that travelled and, more broadly, a research agenda which formed part of the biography of a firm whose main function was to manufacture antibiotics. This particular case of research by an industrial firm was a highly unusual task in Spain in the early 1950s. By the end of the first decade of Franco's dictatorship, contracts for industrial production were starting to be purchased by Spanish entrepreneurs, companies and even the government, so as to participate in the international industrial development that was displaying such promise. Turnkey contracts to establish factories and manufacture a wide variety of products were signed, in what can be regarded as a reaction against the previously compulsory autarchy. When, after a long, almost complete isolation in regard to the industrial means and methods of production, the authorities permitted foreign business relations, political authorities as well as private entrepreneurs looked abroad for industrial manufacturing partners, hoping to join in the success of the *Wirtschaftswunder*. By the early part of that decade, the national borders were being crossed by foreign companies eager to enter new markets, while Spanish entrepreneurs, engineers and scientists had also begun to travel abroad, particularly to the USA, for training in new techniques.[60] Antibiotic screening at CEPA can be regarded as a turn-key research programme, which contributed to learning and training in a factory, both of factory workers and the personnel using microbiological techniques for industrial purposes. Production of other new antibiotics arrived at the same time, and by the mid-1950s streptomycin was also being manufactured through an agreement with Merck.

The screening programme cannot be understood without the factory. Even the agreement between CEPA and Merck can only be comprehended by recalling that the previous agreement for antibiotic manufacturing in Madrid was already underway. Screening became a way of systematically searching for something, whether the presence or absence of a given product, a molecule, enzyme or antibiotic, in a given set of samples, of soil, blood, or a mixture of whatever origin.[61] This way of working had started at Rutgers in the late 1930s.[62] Waksman himself was proud of having led a group of assistants and students who tested more than 10,000 cultures from various natural substrates, 10% of which displayed antibacterial activity; from these, ten antibiotics were submitted for biological evaluation.

The industrial nature of CEPA's research laboratory and, by extension, of the screening programme, was embedded in, or reverted to, natural history-style collection methods.[63] By systematically and manually collecting so much information about different cultures, the programme integrated not only artisanship in every step but also, as microbiological collections do, brought natural history to twentieth-century patterns of industrialising drugs research. The early meaning of the term screening as a microbiological task included a set of artisanal steps, a method of manually collecting data about *Actinomycetes*, and an interaction with the factory system to produce new antibiotics. Women's participation in these tasks has been retrieved from their usual absence in historical reconstructions of drug research and manufacturing. In the CEPA bottling plant, women were the principal workers, with men entering the factory line when large fermentators were set up to produce penicillin.

This also happened with research. A small team of researchers, composed of three graduates including a young female researcher, would build a relationship with scientific researchers at Merck's base in Rahway. Sagrario Mochales and, from the mid-1960s on, Isabel Martín were part of a productive group of scientists that participated in the journeys of knowledge and practices between Merck in Rahway and CEPA in Madrid. New techniques brought from Rahway by Hernandez and Mochales to improve screening results will be discussed in the following chapter.

Notes

1. I am grateful to Sagrario Mochales and Isabel Martín for the information provided in two interviews, in 2002 and 2012, about their careers in CEPA and Merck.
2. Isabel del Cura and Rafael Huertas, *Alimentación y enfermedad en tiempos de hambre. España, 1937–1947* (Madrid: CSIC, 2007); Antonio Cazorla, *Políticas de la Victoria: la consolidación del Nuevo Estado franquista (1938–1953)* (Madrid: Marcial Pons, 2000); Jordi Catalán, *La economía española y la segunda guerra mundial* (Barcelona: Ariel, 1995); Lino Camprubí, 'One grain, one nation: Rice genetics and the corporate state in early Francoist Spain (1939–1952)', *Historical Studies in the Natural Sciences* 40 (2010): 499–531.
3. I have explored research activity in Spain as a combination of national policies and foreign collaboration and influence elsewhere: M. J. Santesmases, 'Severo Ochoa and the biomedical sciences in Spain under Franco, 1959–1975', *Isis* 91 (2000): 706–34; M. J. Santesmases, 'Orígenes internacionales de la política científica', in *Cien años de política científica en España*, eds. Ana Romero and M. J. Santesmases (Madrid: Fundación Banco Exterior, 2008), 293–327.
4. On the concept of screening, see Jean-Paul Gaudillière, 'The singular fate of industrial screening in twentieth century pharmacy: Some thoughts about drugs standardisation and drug regulation', in *Harmonizing Drugs: Standards in 20th Century Pharmaceuticals*, eds. C. Bonah, C. Masutti, Anne Rasmussen and J. Simon (Paris: Glyphe, 2009), 153–80.
5. On antibiotic production and clinical trials, in addition to those cited, see also Harry M. Marks, *The Progress of Experiment: Science and Therapeutic Reform in the United States, 1900–1990* (Cambridge: Cambridge University Press, 2000).
6. The coining of the term, later than that of penicillin, is attributed to Selman Waksman; see S. A. Waksman, 'What is an antibiotic or an antibiotic

substance?' *Mycologia* 39 (1947): 565–9. On antibiotic as a brand, see Robert Bud, 'Antibiotics. From germophobia to the carefree life and back again: The lifecycle of the antibiotic brand', in *Medicating Modern America: Prescription Drugs in History*, eds. Andrea Tone and Elizabeth S. Watkins (New York: New York University Press, 2007), 17–41. On the history of sulfa-drugs, see John Lesch, *The First Miracle Drugs. How Sulfa-drugs Transformed Medicine* (Oxford and New York: Oxford University Press, 2007).

7. See John Krige, *American Hegemony and the Postwar Reconstruction of Science in Europe* (Cambridge, MA: MIT Press, 2006); Benoit Godin, 'Outline for a history of scientific measurement', *Science, Technology and Human Values* 29 (2002): 3–29; M. J. Santesmases, 'Productividad, diplomacia y dependencia tecnica: España en Occidente durante el franquismo', *Pensamiento Iberoamericano* 5 (2009): 21–41.
8. Sławomir Łotysz, 'Controlling the production and distribution of drugs in Communist Poland', *Medicina nei Secoli* 26 (2014): 537–74.
9. Iris Borowy, 'Global health and development: Conceptualizing health between economic growth and environmental sustainability', *Journal of the History of Medicine and Allied Sciences* 68 (2013): 451–85. On technology and post-Cold War ideology and technical performances, see Gabrielle Hecht, ed., *Entangled Geographies: Empire and Technopolitics in the Global Cold War* (Cambridge: MIT Press, 2011).
10. On standards at WHO, see Christoph Gradmann, 'Sensitive matters: The World Health Organisation and antibiotic resistance testing, 1945–1975', *Social History of Medicine* 26 (2013): 555–74.
11. On the linear model, from research to industrial production, see the contributions to part I in Karl Grandin, Nina Wormbs and Sven Widmalm, eds., *The Science-Industry Nexus. History, Policy, Implications* (Sagamore Beach, MA: Science History Publication, 2004). See also Marion A. Hulverscheidt, 'The scientific entrepreneur or financing in pharmaceutical research. A portrait of the mariologist Werner Schulemann, 1888–1975', in *Perspectives on Twentieth Century Pharmaceuticals*, eds. Vivian Quirke and Judy Slinn (Oxford: Peter Lang; 2010), 121–48. This is an account of research carried out by a scientist who began his career at an industrial research laboratory.
12. Robert F. Bud, 'Strategy in American cancer research after World War II: A case study', *Social Studies of Science* 8 (1978): 425–59.
13. See the contributions to Ilana Löwy and Jean-Paul Gaudillière, eds., *Invisible Industrialist. Manufactures and the Production of Scientific Knowledge* (London: Macmillan, 1998).
14. See William Kingston, 'Streptomycin, Schatz v. Waksman, and the balance of credit for discovery', *Journal of the History of Medicine and Allied*

Sciences 59 (2004): 441–62, where it is claimed emphasis was placed on the repetitive nature of screening to reinforce Waksman's authority and thus wrest authority from his colleagues, particularly Albert Schatz. On this see also Milton Wainwright, 'Streptomycin: Discovery and resultant controversy', *History and Philosophy of the Life Sciences* 13 (1991): 97–124. The testimony of Schatz himself can be found in Albert Schatz, 'Some personal reflections on the discovery of streptomycin', *Pakistan Dental Review*, 15 (1965): 125–34.

15. On technology transfer from abroad into Spain see Mar Cebrián, 'Technological Imitation and Economic Growth during the Golden Age in Spain: 1959–1973' (PhD diss., European University Institute, Florence, 2004); Santiago López and Mar Cebrián, 'Economic growth, technology transfer, and convergence in Spain, 1960–1973', in *Technology and Human Capital in Historical Perspective*, eds. Jonas Ljunberg and Jan-Pieter Smits (New York: Palgrave-Macmillan, 2004), 120–44. On the factory system of production from the perspective of the history of economics and industry see Stanley D. Chapman, 'The textile factory before Arkwright: A typology of factory development', *The Business History Review* 48 (1974): 451–78; Bo Carlsson, 'The development and use of machine tools in historical perspective', *Journal of Economic Behaviour and Organisation* 51 (1984): 91–114; Joel Mokyr, 'The rise and fall of the factory system: Technology, firms and households since the industrial revolution', *Carnegie-Rochester Conference Series on Public Policy* 55 (2001): 1–45.
16. On resistances as part of this story of penicillins in Spain see Chap. 7. On resistances in the USA, Scott H. Podolsky, *The Antibiotic Era: Reform, Resistance, and the Pursuit of a Rational Therapeutics* (Baltimore: Johns Hopkins University Press, 2014). For a history of resistances *avant la lettre*, see Christoph Gradmann, 'Magic bullets and moving targets: Antibiotic resistance and experimental chemotherapy, 1900–1940', *Dynamis* 31 (2011): 29–46.
17. H. B. Woodruff and L. E. MacDaniel, 'The antibiotic approach', in *The Strategy of Chemistry*, eds. S. T. Cohen and R. Rowatt (Cambridge: Cambridge University Press, 1958), 29–48, on 30.
18. Woodruff and MacDaniel, 'The antibiotic approach'.
19. Woodruff and MacDaniel, 'The antibiotic approach'. On the US pharmaceutical industry and antibiotic production (and the impact of resistance), see Podolsky's *Antibiotic Era*, Chaps. 1, 2 and 5. On spaces for industry opened up by resistance, see Gradmann's 'Magic bullets'. Also Arthur Daemmrich, 'Synthesis by microbes or chemists? Pharmaceutical research and manufacturing in the antibiotic era', *History and Technology* 25 (2009): 237–56.

20. H. B. Woodruff, Oral history interview, 18 August 2004, by Shaun Illingworth and Nicholas Trajano Molnar, p. 36, Rutgers Oral History Archives, at http://oralhistory.rutgers.edu/rutgers-history/31-interviewees/1337-woodruff-h-boyd, accessed June 28, 2015.
21. María Isabel Porras Gallo, 'Antecedentes y creación del Instituto de Sueroterapia, Vacunación y Barteriología de Alfonso XIII', *Dynamis* 18 (1998): 81–105.
22. Nuria Puig, 'El crecimiento asistido de la industria química en España: Fabricación Nacional de Colorantes y Explosivos, 1922–1965', *Revista de Historia Industrial* 15 (1999): 105–36; Nuria Puig and Javier Loscertales, 'Las estrategias de crecimiento de la industria química alemana en España, 1880–1936: exportación e inversión directa', *Revista de Historia Económica* 19 (2001): 345–82.
23. Lesch, *The First Miracle Drugs.*
24. On culturing techniques for human tissues, see Hanna Landecker, *Culturing Life: How Cells Became Technologies* (Cambridge, MA: Harvard University Press, 2007).
25. Santesmases, 'Productividad, diplomacia y dependencia técnica'.
26. Although a discussion of the similarities and differences between these two kinds of screening is beyond the scope of this book, I am grateful for comments on this issue from a reviewer of an early version of the text. On the screening of viral particles for the treatment of human cancer in the 1950s, see Jean-Paul Gaudillière, 'Circulating mice and viruses: The Jackson Memorial Laboratory, the Cancer Institute and the genetics of breast cancer, 1930–1965', in *The Practices of Human Genetics*, eds. Michael Fortun and Everett Mendelsohn (Dordrecht–Boston: Kluwer, 1999). On the screening of cervical cancer, see Ilana Löwy, *Preventive Strikes. Women, Precancer and Prophylactic Surgery* (Baltimore: Johns Hopkins University Press, 2010). On PKU, see D. B. Paul and J. P. Brosco, *The PKU Paradox: A Short History of a Genetic Disease* (Baltimore: Johns Hopkins University Press, 2013).
27. On the factory system in historical perspective see references in note 15.
28. See Robert Bud, *The Uses of Life* (Cambridge: Cambridge University Press, 1993).
29. John P. Swann, *Academic Scientists and the Pharmaceutical Industry: Cooperative Research in Twentieth-century America* (Baltimore: Johns Hopkins University Press, 1988), 65–6.
30. Lesch, *The First Miracle Drugs*, 198–202.
31. Quoted in John Parascandola, 'The introduction of antibiotics into therapeutics', in his *Studies in the History of Modern Pharmaceutical and Drug Therapy* (Farham: Ashgate 2012), 261–80, on 262. See also S. A. Waksman and H. Boyd Woodruff, 'The soil as a source of

microorganisms antagonistic to disease-producing bacteria', *Journal of Bacteriology* 40 (1940): 581–600; S. A. Waksman, 'Antagonistic relations of microorganisms', *Bacteriology Reviews* 5 (1941): 231–91.
32. Dayle A. Harris and Myrle L. Ruger, 'Microbiological aspects of new antibiotic screening. A plug test procedure', *Antibiotics and Chemotherapy* 3 (1953): 265–70.
33. Hubert Lechevalier, 'The search for antibiotics at Rutgers University', in *The History of Antibiotics. A Symposium*, ed. John Parascandola (Madison: American Institute of the History of Pharmacy, 1980), 113–23.
34. H. Boyd Woodruff and Selman Waksman, 'Historical background', *Annals of the New York Academy of Sciences* 89 (1960): 287–98.
35. Alex Sakula, 'Selman Waksman 1888–1973, discoverer of streptomycin: A centenary review', *British Journal of Diseases of the Chest* 82 (1988): 23–31.
36. Swann, *Academic Scientists*, 87–90.
37. Selman Waksman, 'Streptomycin: Background, isolation, properties, and utilization', *Nobel Lecture* (December 12, 1952), 370–88, at http://nobelprize.org/nobel_prizes/medicine/laureates/1952/waksman-lecture.pdf.
38. Accounts of these events are in Wainwright, 'Streptomycin'.
39. A list of diseases responding to streptomycin can be found in Waksman, *Nobel Lecture*, 383–4. The activity of streptomycin in the treatment of tuberculosis was widely celebrated at the time. See Julius H. Comroe, 'Pay dirt: The story of streptomycin part II', *American Review of Respiratory Disease* 117 (1978): 957–68.
40. Lechevalier, 'The search for antibiotics'.
41. Swann, *Academic Scientists*, 90.
42. H. Boyd Woodruff, 'Natural products from microorganism', *Science* 208 (1980): 1225–9.
43. Waksman, *Nobel Lecture*.
44. In 1952, 44 antibiotics were listed by William Spector, ed., *Handbook of Biological Data* (Philadelphia and London: W.S. Saunders Co., 1952) and Lechevalier, 'The search for antibiotics'.
45. Waksman, 'What is an antibiotic'; Bud, 'From germfobia to carefree life'.
46. W. R. Strohl, H. B. Woodruff, R. L. Monaghan, D. Hendlin, S. Mochales, A. L. Demain and J. Liesch, 'The history of natural products research at Merck & Co.', *SIM News* 51 (2001): 5–19.
47. Swann, *Academic Scientists*.
48. Fondos Servicio de Estudios del Banco Urquijo, Instituto Juan March de Investigaciones Sociales Avanzadas, Madrid. I would like to thank Antonio Gómez Mendoza for his advice in consulting this archive.
49. 'Agreement made and entered into as of the 14th of January 1949 between Merck & Co., Inc and Banco Urquijo', contracts 37QC, boxes 5112 and 5113, IDD 1.04. Ministerio de Industria. Sección de Subsecretaría

(Asistencia Técnica), Archivo General de la Administración (AGA), Alcalá de Henares, Madrid (in English in the original). I would like to thank Mar Cebrián for her guidance in consulting these files at the AGA.

50. Sagrario Mochales, interview with the author.
51. Strohl et al., 'The history of natural products', 9.
52. Woodruff, n. 19: 23.
53. Strohl et al., 'The history of natural products'.
54. Sagrario Mochales, interview with the author, 2012. See also Sagrario Mochales, 'Ten years of CIBE symposia, 1989–1998', *International Microbiology* 1 (1998): 251–4.
55. 'Agreement made and entered into as of the 14th of January 1949 between Merck & Co., Inc and Banco Urquijo'.
56. On PKU, D. B. Paul and J. P. Brosco, *The PKU Paradox: A Short History of a Genetic Disease* (Baltimore: Johns Hopkins University Press, 2013). On screening of cancer viruses see Löwy and Gaudillière, *Invisible Industrialist*, and Gaudillière, 'Circulating mice and viruses'. On cancer therapy, Bud, 'Strategy in American cancer research'. On sulfa-drugs, see Lesch, *The First Miracle Drugs.*
57. Sagrario Mochales, interview with the author, 2012.
58. S. Hernández, Sagrario Mochales, J. Martínez Mata and A. Gallego, 'Un programa de investigación de nuevos antibióticos. I. Resultados del año 1957', *Anales del Instituto de Farmacología Española* (1958): 263–6.
59. Paul and Brosco, *The PKU Paradox.*
60. See Lino Camprubí, *Engineers and the Making of the Francoist Regime* (Cambridge, MA: MIT Press, 2014). A Spanish version has just been published with additional chapters and a greater connection to the Spanish historiography: Lino Camprubí, *Los ingenieros de Franco: Ciencia, catolicismo y Guerra Fría en el Estado franquista* (Barcelona: Crítica, 2017).
61. For a broad discussion on screening in relation to standards and regulation see Jean-Paul Gaudillière, 'The singular fate'.
62. Selman Waksman, 'The antibiotic era: A history of the antibiotics and of their role in the conquest of infectious diseases', published posthumously, quoted in J. H. Comroe, Jr., 'Pay dirt: The story of streptomycin. I: From Waksman to Waksman. Retrospectroscope', *American Review of Respiratory Diseases* 117 (1978): 773–81.
63. Collecting data is suggested as a natural history practice by Bruno Strasser, 'Collecting and experimenting: The moral economies of biological research, 1960–1980s', in *History and Epistemology of Molecular Biology and Beyond: Problems and Perspectives* (Berlin: Max Planck Institut für Wissenschaftsgeschichte, 2006), preprint 310, 105–25. Also Bruno Strasser, 'Collecting nature: Practices, styles, and narratives', *Osiris* 27 (2012): 303–40.

CHAPTER 6

A Promising Drug: Bacteria, Antibiotics and Marketing in an Era of Economic Development

On February 15, 1991, the weekly pages devoted to medicine and health in the Barcelona daily newspaper, *La Vanguardia*, included a brief report on the effects of the antibiotic fosfomycin in the treatment of urinary infections. When administered as a salt, the report claimed, the antibiotic cured the patient in just one day.[1] Manufactured in Spain since the early 1970s, fosfomycin originated from the circulation of people and objects during the economic miracle of the 1960s in the West. The antibiotic activity had been detected in 1966 by a scientist at the research laboratory of the Spanish pharmaceutical firm CEPA (Compañía Española de Penicilina y Antibióticos), established through an agreement with the US firm Merck signed in 1954 (see Chap. 5). After eight years of systematically testing hundreds of samples, a new screening method led to the detection of an unknown antibiotic. This discovery has been proudly recalled in accounts by the participating Spanish researchers and in the minutes of CEPA's annual board meetings at the time.[2] The new antibiotic isolated and identified as part of the research programme was named phosphonomycin, later changed to fosfomycin and finally, for marketing within Spain, *fosfocina*. In this chapter I relate the story of an expectation fulfilled by the production of a new drug, developed in the factory, the clinic and the research laboratory. The long life of this antibiotic, later retrieved and promoted for the treatment of urinary infections as if a recent discovery, began as the

M. J. Santesmases, *The Circulation of Penicillin in Spain*,
Medicine and Biomedical Sciences in Modern History,
https://doi.org/10.1007/978-3-319-69718-5_6

result of the new methods of detecting antimicrobial activity introduced in the early 1960s precisely as an outcome of the research on the action mechanism of penicillin in bacteria, transferred from a US laboratory to a Spanish pharmaceutical firm.[3]

In the case of *fosfocina*, exploration of mode of action became embedded in the research programme encompassing the race for new antibiotics. The effects of penicillin on bacteria had been studied during the war, not only in the hunt for antimicrobial activity but also as a tool in research on the composition and structure of the bacterial cell wall, and in a combination of previous approaches to the cell envelope in the biochemistry of phage infection. Penicillin was tested on infections but, at the cellular and molecular level, its action was unknown during its early clinical era. Invisible as the bacterial wall was for clinical purposes, in the biological laboratory the cellular meaning of penicillin action made the bacterial wall visible through both microbiology and biochemistry. Penicillin, and later other antibiotics, acquired a social life beyond the clinic. As a medical commodity originating from the factory, it was a combination of biological approaches that turned this new drug—a medical and industrial product in itself—into an agent in the era of biomedical research.[4] The proliferation of descriptive titles conferred on the post-war and Cold War era—the atomic era; the era of biomedicine; the antibiotic era—illustrates the overlapping strategies based in science; a set of terms denoting the practice of experimentation in the standard space that the research laboratory was becoming in atomic sciences, drug development and medicine.

According to post-war ideology concerning infection cure and the promises of scientific research, technology would eventually defy resistances. The emergence of resistant bacteria added impetus to the search for new antibiotics, while resistance as a biological response to antibiotic action became an integral part of the research programme.[5] The agenda for microbiology and the pharmaceutical industry was still to find new therapeutic agents that would demonstrate their healing capacity against many infections. During the 1950s, the isolation of ten novel compounds resulted from the testing of between 10,000 and 20,000 cultures. Over time, as knowledge of compounds increased, the number of new antibiotics identified and isolated decreased.[6] New screening strategies were required.

An Interactive Research Agenda

The laboratory experience was able to detect only parts of complex natural phenomena; available antibiotics were therefore added to eliminate susceptible substances, aiding the isolation of new microorganisms, from which new antibiotics could be produced, according to a report on the method by a group of Merck researchers.[7]

The soil sample would be cleaned, 'creating a selective environment of the desired ones by the addition of antibiotics to the plating media'. The culturing media manufactured thus generated the knowledge and practices to be mobilised in order to find other, new antimicrobials. The Petri dish—assisted by soft artefacts, such as paper disks soaked with antibiotics placed on the surface of the culture medium—was the laboratory container of these quests for new antibiotic activity. To distinguish any new compound with antibacterial activity, a set of bacteria isolates had been selected on the basis of their resistance to available antibiotics.[8] The set would remain as a permanent reference of the need to search for new therapeutic substances to fight against, mostly, Gram-negative substances.

The steps included in the programme were a combination of microbiological practices embedded in an industrial project. The CEPA–Merck screening programme was set up in a factory with an industrial agenda and a demand for high productivity. Merck provided the group with appropriate training and tools and CEPA researchers were aware of the aims: industrial manufacturing of a product, a new therapy against infection.

As detailed in the previous chapter, Antonio Gallego, who had been the link with Merck since the early days of CEPA, took charge of the initiative. The head of the quality control department, Justo Martínez Mata, was also appointed head of the research laboratory. Time and again, however, the same, already known antibiotics were discovered: different microorganisms could produce the same substance.[9] 'At this point', it was decided, 'a more creative research approach to screening becomes essential.'[10]

Constructing the Target: The Bacterial Cell Wall

In 1962, at Merck's Rahway laboratories, Eugene Dulaney established a new antibacterial detection assay. Known as the spheroplast method, it was based on the physical effects produced by antibiotics in the cell walls of bacteria. Although Dulaney never published on his achievements, one of his former colleagues at Merck, Lynn Silver, later emphasised his role in developing 'the art and science of antibacterial screening'. As Silver notes, Merck scientists only published this method in 1992.[11] Indeed, it was considered more instrumental—and thus more in need of confidentiality—than those methods used to manufacture antibiotics in the factories, which were patented and thus became publicly documented.[12]

Dulaney's method was praised by scientists at both the Merck and CEPA laboratories: microorganisms active against bacteria would now be found in soil samples 'from all over the world'. It appears the particular procedure of antimicrobial detection and identification was considered more crucial than the actual samples: activity could be overlooked by failure to use 'adequate' methods. It was concluded there was 'no perfect screening system … each change in method will lead to a corresponding change in results'.[13]

I will now narrate a brief history of the bacterial cell wall, from the early 1940s until the 1960s, to situate the method that led to the detection of a new antibiotic in Madrid in 1966. Through this time the social life of antibiotics retained agency in the research laboratory, in an epistemic space that would become known as cell biology, and participated in the origins of the academic space of molecular biology.[14] In both timelines penicillin contributed to the knowledge and practices surrounding the bacterial cell wall. While being used as a tool in the research laboratory it would eventually participate in the finding of other, new antimicrobials. This biological object the bacterial wall became belonged to the epistemology of the antibiotic era, in which an industrial product intervened in biological and biomedical research. This interaction was a new platform both for the industrial production of antibiotics and for biology, producing new knowledge about bacteria—about its wall, its morphology and chemical composition—and new antibiotics. *Fosfocina* was the first; others that followed include cephamycin, thienamycin, carbapenems (epithienamycins) and uracil-containing antibiotics.[15]

The clinic, and therefore medical practice, had its own particular trajectory and agenda during the early days of penicillin: a commitment to testing the drug and establishing protocols for its use against infections. The rapidly apparent resistances to penicillin—and, slightly later, many other antibiotics—were observed at the bedside.[16] The laboratory bench remained a parallel testing site, however: penicillin—as has been the case with many drugs—was used in medical practice due to knowledge of its effect on bodies, while its activity at the level of cells and biological molecules was still unknown.[17]

As early as 1940, Arthur D. Gardner, a member of Howard Florey and Ernest Chains's penicillin study group at the William Dunn School of Pathology in Oxford, had noticed a morphological effect, a change in the appearance of the growth of *Cl. welchii* in low concentrations of the drug. These low concentrations—less than was required for a 'full inhibiting' of bacterial growth—produced an elongation of the cells. He also reported that *Staphylococcus aureus*, when grown in the presence of penicillin produced by *Penicillium notatum*, underwent spherical enlargements and was subject to 'imperfect fission'. Gram-negative bacteria that were resistant to penicillin displayed this shape-changing phenomenon 'very well'.[18]

Luis Despain Smith and Telma Hay, at the Biochemical Research Foundation of the Franklin Institute in Delaware, suggested 'it seems possible that penicillin either has some action on the cellular wall of *S. aureus* or that it interferes with the assimilation of one or more growth factors necessary for the actual fission of the growing cell'.[19] By introducing the phenomenon of interaction between the bacterial cell wall and penicillin, they positioned the wall as the part of the bacteria that could be targeted by the new drug. In the presence of small amounts of penicillin, bacteria underwent lysis and released protoplasmatic substances into the culturing medium, detectable to the naked eye through the resulting turbidity. Stained by gentian violet, the culture revealed imperfect fission in the photomicrographs that darkened as the bacteria expanded: darkness that signified morphological changes.

At the Biochemistry Department at Cambridge University, Ernest Gale had been impressed by the effects of penicillin on staphylococcal infections during the Second World War, and began studying the mode of action of penicillin, and later other antibiotics, on Gram-positive and Gram-negative bacteria following the war.[20] By keeping the biochemistry of microorganisms and antibiotics at the core of his experimental skills and research agenda, he followed the biochemist Marjory Stephenson—to whose chair at Cambridge, Gale would later be appointed—in her 'use of the microbe as experimental material'.[21]

The Australian Milton Salton, having completed his PhD under Gale, became an expert in isolating bacterial walls for effective electronic microscopy. In the early 1950s, while at the department of bacteriology of the University of California, Berkeley, Salton shifted to biochemistry. Still focusing on bacterial cell wall images, he declared the effects of some enzymes to be 'the most elegant method of studying the nature and location' of the bacterial cell wall and its components.[22] Treatment of a set of Gram-negative and Gram-positive bacteria resistant to lysis by lysozyme and trypsin enabled him to characterise the set and offer 'some indications of the suitability of using heat-killed bacteria for the isolation of cell-wall decomposing microorganisms'[23]; that is, to participate in the search for new antimicrobials. He provided images of bacterial cell walls as evidence of his experiments, and of the resistance to digestion found in some enzymes of Gram-negative bacteria. Salton concluded that the resistance of heated Gram-positive bacteria to lysis 'is largely governed by the nature of the cell walls' and suggested a method for detecting microorganism activity 'capable of producing cell wall decomposing enzymes'.[24] Decomposed cells were illustrated by the wall's disappearance, 'digested' through a combination of heat and lysozyme action. Lytic substances were made visible precisely through their action—the action sought in the hunt for new antimicrobials—at a time when Salton's research in the USA was funded by a Merck International Fellowship.

By the early 1950s, cellular units of Gram-positive microorganisms that had lost their rigid cell wall—that is, non-animal cells—had become known as bacterial protoplasts. The term came from botany: the living components of plant cells were known to have a membrane, but not a wall. The rigidity of that which characterised bacteria differed in permeability: the wall was a barrier while the membrane was a *passage* tissue.[25] Both the wall and the membrane contributed to an understanding of cell function and behaviour, becoming mutual references, not in comparison with one another but as aids to understanding their respective biologies: as one German plant biologist phrased it, their metabolism, biosynthesis and energy transfer.[26]

US bacterial geneticist Joshua Lederberg, then at the University of Wisconsin-Madison, identified protoplasts in bacteria as cells with their wall removed, and devised a method for obtaining them through the action of penicillin.[27] Without the wall, the cell was transformed into a 'spherical protoplast' and its sensitivity to a hypotonic medium produced lysis. After two hours under the action of a penicillin solution in the presence of magnesium salt, Lederberg observed the cells of *Bacilus magaterium* and *Escherichia coli* take on a spherical shape and, once water

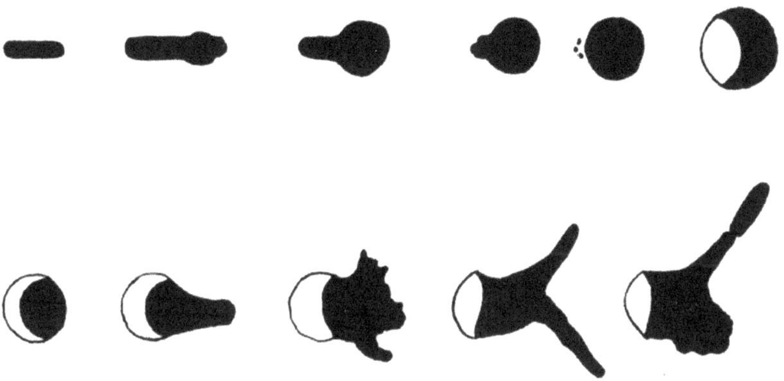

FIG. 1.—*Above:* Transformation of bacterial rod to spherical protoplast in the presence of penicillin. *Below:* reversion of protoplast to rod in absence of penicillin. Observed under dark phase contrast.

Fig. 6.1 The transformation of *E. coli. Source:* Joshua Lederberg, 'Bacterial protoplasts induced by penicillin', *Proceedings of the National Academy of Sciences of the United States of America* 42 (1956): 575

was added, break down (Fig. 6.1). These were the two features of protoplasts, a spherical shape and lysis in distilled water: in the absence of hypotonic media, protoplast suspension was able to 'remain intact' for several days at 4°C.

For his 1958 Nobel lecture, Lederberg discussed his experiments with both streptomycin and penicillin, neither of which produced 'direct mutation'.[28] Drug resistance had become a significant issue by that time, at least among some biomedical researchers in the USA, Norway and Japan, and would continue to increase in significance, intensifying warnings, research interest and funding to combat resistance as a phenomenon, produced by drug use in both the microbiological laboratory and the clinic.[29] But according to Lederberg's interpretation, presented in Stockholm in May 1959, the cell was killed by these substances before they accessed chromosomes.

Together with Jacqueline St Clair, Lederberg had by then solved the problem of bacterial shape-changing through environmental action. They had explored an environment composed of penicillin, magnesium salts and water, and the extent to which protoplasts behaved as cells or not, according to the amount of penicillin in the culture media. As Lederberg's research agenda was bacterial genetics, his main interest was mutation, which did not occur in the experiments he performed with St Clair. Their results were clearly summarised: 'the mechanism of action of penicillin is to inhibit the synthesis of the bacterial cell wall'.[30] By inhibiting bacterial growth without affecting human cells, penicillin appeared to recover its benefits for the clinic: by inhibiting one type of cells—bacterial—it stabilised its ability to protect others—those of the human body—from such bacteria.

At the Biochemistry department of the College of Agriculture at the University of Wisconsin, James T. (Ted) Park isolated from penicillin-inhibited *staphylococci* a very unusual nucleotide containing uridine diphosphate (Park and Johnson 1949). In the collaboration between Park and Jack Strominger, the action of penicillin—its 'selective toxicity'—was explained as an 'interference with the metabolic sequence of the biosynthesis of the cell wall, those metabolic sequences that are not found in animal cells but indeed were in bacteria'. A cell-wall research community was being created at this time, as illustrated by the letters exchanged between Park, Strominger and Lederberg,[31] who all published on the mechanisms of penicillin action in 1957, Lederberg by himself in a short note for the *Journal of Bacteriology*, Park and Strominger with a longer paper.[32] Competing and collaborating at the same time, Lederberg proposed a submission to the same journal but Park argued that his paper with Strominger, being longer, was better suited for *Science*.[33]

The wall was a collective scientific object in motion, the circulation of which as a term included the dissemination of images, practices and knowledge regarding the composition of the wall, its process of destruction and the action of penicillin on the many bacterial cell walls investigated at the time.

Bacteria Cells and the Negotiation of Terminology

Since the acceptance of cell theory, the cell as a biological object had operated as an articulator of the field, enabling 'biology' to secure disciplinary, academic space. It would incorporate natural history, the natural sciences and bacteriology when dealing with the form and function of living beings, studied at the level of their basic structured components: the cells. Cytology would develop techniques for observing cell components as biological material; the nucleus, cytoplasm and a number of organelles would be displayed for the microscope, through processes involving not only staining—this pertained to the cell as a whole, and therefore, to histology—but destruction of the membrane.[34]

The chemistry of such components, the kind of substances that constituted the membrane and, in the case of bacteria, the wall, remained unknown. In this process, as shown in Fig. 6.1, penicillin as a wall-breaker became a tool for determining the chemical composition of the bacterial cell wall. By operating as a laboratory substance, penicillin contributed to unveiling its own function at the cell level. This life at the research bench, later shared by other antibiotics, exhibits their long-lasting function beyond the medical bedside.

Research into penicillin's action became a transnational project. Initially led by the USA and Britain, it became international with the involvement of other researchers who, from chemistry, microbiology and the new space of molecular biology, explored the molecules involved in the life of bacterial cells. Throughout this process, new names and new meanings were negotiated as they became part of the story.

This was certainly the case for the terms protoplast and spheroplast, mobilised by bacteriologists and biochemists from 1955 onwards.[35] Exchanges of letters and references in papers suggest that negotiating the use of these terms also involved negotiation of biological meanings, of representations of the bacterial wall and its biological function: that is, reaching agreement on, among other things, what the bacterial cell wall was, its composition, chemistry, interactions with phages, morphology and osmotic sensitivity.

A pioneer of protoplast studies, Kenneth McQuillen from the department of biochemistry at the University of Cambridge, actively negotiated the meaning of the term protoplast. McQuillen discussed with Lederberg the convenience of defining its meaning as a bacterial cell that had lost its cell wall. McQuillen had described protoplasts in collaboration with Salton at Cambridge, which situated him in the earliest cell wall research network of the 1950s.[36] He showed protoplasts from *Bacillus megaterium* dividing under particular conditions (see Fig. 6.2).[37]

The text McQuillen sent to Lederberg for approval—also circulated by the Swiss electron microscopist, Edward Kellenberger—began with a declaration of what the bacterial cell wall was composed of, and cited studies by the signatories and others. The list of authors illustrates an international consensus: Swedish bacteriologist, Claes Weinbul; German virologist, Wolfhard Weidel; Lederberg; Salton; Kellenberg; McQuillen himself; German biologist, Emmy Klieneberger-Nobel—the only woman

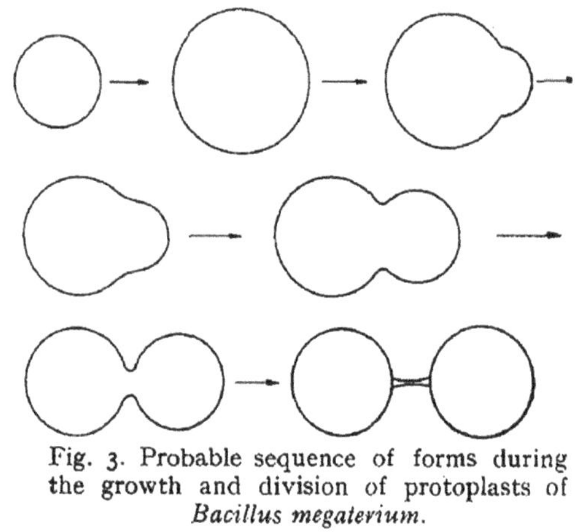

Fig. 6.2 Protoplasts from *Bacillus megaterium*. *Source:* Kenneth McQuillen, 'Bacterial protoplasts: Growth and division of protoplasts of *Bacillus megaterium*', *Biochimica et Biophysica Acta* 18 (1955): 461. Reproduced with permission

on the list—who had worked at the Lister Institute since the Second World War; Spanish electronic microscopist, Miguel Rubio-Huertos; British microbiologist R. E. Strange from the Microbiological Research Station (UK Ministry of Supply); and J. Tomscik from the University of Basel. It was written as a scientific letter to the editor—a short article—with 22 footnotes. After briefly presenting the knowledge and methods available on the isolation and composition of the wall, the text stated the need to avoid 'a situation in which the word PROTOPLAST in one context means that part of a bacterial cell which lies within the cell wall and in another context means something different' and proposed the term be solely used 'to describe a structure in which the cell wall is known to be absent'.[38]

On receiving a draft of the letter, Lederberg deemed McQuillen's 'question about terminology' to be 'rhetorical'. After discussing McQuillen's interest in constructing a shared meaning of the term, he added, 'I would rather fear that if say "spheroplast" were adopted it would come to connote the alternative class of false protoplast rather than the inclusive class that needed to be cited'. He regarded the criticism about the characterisation of protoplasts 'entirely valid' but wondered whether it merited an approach from 'the semantic side'.[39]

Closer to biochemistry than the geneticist Lederberg, McQuillen was unable to secure Lederberg's support. He was also unable to include Weidel: 'the only dissentient', as McQuillen phrased it to Lederberg, among those who received a draft of the letter. From the Max Planck Institute of Biology at Tübingen, Weidel wrote, in a rather derisory tone, 'even a megaterium protoplast is not a protoplast in the old sense ... I should call this envelope the second innermost layer of the wall'.[40]

Circulating terms, even if to challenge previous meanings, afforded them reliability. The tones and arguments invoked to convince, or resist acceptance engaged in the contest of scientific authority and academic recognition, an interchange of respect through the networking practices of an emerging community of cell biologists. The term spheroplast was already in use when a collective statement on the wording 'bacterial protoplast' was finally published in 1958, signed by 13 researchers from 12 institutions and six countries.[41] McQuillen further developed the meaning and normativity of the cell at the International Congress of Biochemistry held in Vienna in 1958, where he summarised the 'comparative biochemistry of bacterial cell walls, protoplasts, and spheroplasts' and used an entire chapter of the resulting publication to insist on the meaning of protoplasts as defended in the letter to *Nature*.[42] Later, Weidel together with Pelzer (1964) described 'the fragile structure underneath the murein skeleton of Gram-negative bacteria' as a 'spheroplast', 'very suitable objects for studying murein metabolism'. For Weidel, the spheroplast, obtained by removing the cell wall, rendered the cell 'mechanically fragile'.

It was the fragility of the bacterial wall that became the focus of CEPA and Merck's screening method. The spheroplast, while being shaped as an international research object was, as a word, involved in the early interchanges about the making of the bacterial cell wall; it was used to name a body without the murein sacullus (coined in Germany by Wolfhard Weidel) or peptidoglycan (coined by Belgian microbiologist Jean-Marie Ghuysen), that is, which has lost its wall.[43] Both names—murein and peptidoglycan—have been in use ever since. As a target, the bacterial wall participated in the origins of cell biology as a research space, in the developments of the biochemistry of the day and also in electron microscopy studies of biological forms composed of large molecules; the wall was described as a 'bag-shaped' molecule by Weidel and Pelzer.[44] It was penicillin as a molecule investigated for medical purposes within microbiology that mobilised this laboratory activity, the experimental practices of which became immediately embedded in industrial research. This feedback loop, although provoked by penicillin, was fed by the walls of infection-inducing bacteria, against which industry was allocating large budgets to construct a culture of scientific and industrial success based on research. While industrial research was serving medicine by manufacturing new drugs, between the factory line and the consulting room, a medical biology—that is, cell biology—was constructing its own space in the biomedical era. The laboratory challenged medical practice while at the same time contributing to its success. The social life of antibiotics was wider than a professional expert space. By the 1960s it was occupying diverse rooms: the factory, the biological laboratory and the clinic, in research, industry and health care.

New Methods for a New Drug

The spheroplast, as a cellular body without a wall, became the biological object sought by the new antibiotic screening method. At the Merck Research Laboratories, Eugene Dulaney helped overcome the stagnation of antibiotic discovery with the previous screening method. By this time, microbiologists had developed complex skills for handling bacterial broths. Usually one bacteria sample would contain more than one antimicrobial, and when tested as an extract against a known set of bacteria, some of the active compounds contained would easily be masked by precisely their own antimicrobial capacities, able to kill some of the unknown bacteria included in the sample. While microbiological skills developed during these years of screening, knowledge about microorganisms increased rapidly, as did the number of cultures registered.[45]

Dulaney's method to detect antimicrobial activity by identifying spheroplasts was applied to partially purified extracts from soil samples. Some hours after being added to a bacterial medium, the shape of the bacteria could be observed though a microscope. A round form was the first indication of antibacterial activity, enlargement proof of the presence of a product acting against the bacterial wall and releasing spheroplasts. The Sphero detection assay was combined with Perm, a colouring test designed by Merck microbiologist David Hendlin. An indicator—a chromogenic substance—was added: whenever the microorganism under study produced a reactive substance, a round coloured line would be generated in the culture. Corroboration of the Sphero test in the CEPA–Merck screening programme provided a zone of blue when observed under ultraviolet light.[46]

The combination of Sphero and Perm methods as screening strategies contributed to the detection of new antimicrobial activity identified in a soil sample taken in Alicante during April 1966; Sagrario Mochales detected a tiny activity in the culturing media the following month. The sample exhibited initial activity by generating tiny spheroplasts, leading to further purification of the agent. Successive Perm procedures revealed the zone of blue. The sample was then tested on a set of resistant bacteria, a checklist used by the Merck and CEPA researchers involved in the screening programme.[47]

The extract identified proved to be active in vitro against a set of bacteria, both Gram-negative and Gram-positive. When tested in mice, the first extracts displayed activity against many experimental infections. This meant the identified substance was more active in vivo than in vitro, a fact that generated considerable interest, prompting the group to conduct further research to identify the substance responsible.[48] The activity was confirmed and a concentrated sample lyophilised for transportation as a solid powder. The active substance was subsequently identified at the Merck Research Laboratories, and isolated as a calcium salt. It proved to be an unusual chemical compound: a cycle of three atoms, one of oxygen, in a phosphonic acid.[49]

According to accounts by the scientists involved, it was the combination of a standardised protocol in the treatment of soil samples and the increasing skills in handling microbes and soil samples among the CEPA group that led to identification of the new drug.[50] Sagrario Mochales remembers slightly modifying the method by keeping the culture longer than indicated, based on the weak activity she detected during the standard time. It was precisely this additional culturing time that allowed the activity to become clearly visible. Even with the low degree of purity of the first extracts, the culture showed high activity against many experimental infections in mice. Careful handling and observation of cultures were instrumental elements of microbiological practice.

Celebrating a Research Success

Detection of the new drug was reported in 1969 at the annual meeting on infectious diseases of the American Society for Microbiology (9th Interscience Conference on Antimicrobial Agents and Chemotherapy).[51] The culture was identified as MA-2898 (and by the Northern Regional Research Laboratory as NRRL-B-3357), as previously described by Waksman.[52] Later, other cultures isolated from soil samples were identified as containing phosphonomycin. The authors reported that oral administration of the new drug to mice experimentally infected with Gram-positive and Gram-negative bacteria, demonstrated activity not only comparable to antibiotics that already existed but also active against some germs resistant to the antibiotics currently in use. Calcium salt taken orally provided an antibacterial level in blood in two hours, and the antibiotic was non-toxic.

In October 1969, the new antibiotic was described to the scientific community as having been isolated from *Streptomyces fradiae*.[53] Published in the US journal *Science*, the letter announcing the discovery had 14 signatories, 12 from Merck Sharp and Dhome research laboratories and three from the CEPA research laboratory: Justo Martínez Mata, director of the CEPA laboratory; Sebastián Hernández, a medical researcher on the screening programme; and Sagrario Mochales, who had identified the activity, modified the method and contributed to its isolation. If, as a Foucauldian 'dispositive', the number of signatures indicates the drug's discovery to be a collective endeavour—note that Dulaney's name is not included, but Hendlin's is—the order of names speaks of managerial practices and hierarchies at the workplace: with her name in the final position, Mochales's authority as the research worker who obtained experimental evidence of the new antibiotic activity was acknowledged.[54]

A second letter to *Science* followed, concerning the isolation, purification and identification of the antibiotic by a group of chemists, this time exclusively from Merck, revealing the absence of chemical skills in the CEPA group. The structure of this collaborative endeavour involving firms from two countries was determined in the Merck research laboratory. The entire trajectory of the research programme exhibited both international collaboration and a hierarchy based on the technical, scientific and industrial authority of Merck, alongside the instrumental participation of a handful of young Spanish researchers.

The discovery of the new antibiotic phosphonomycin was reported in the Spanish medical weekly *Noticias Médicas* a few days after these letters were published in *Science*.[55] In a full-page article, Antonio Gallego presented phosphonomycin to the Spanish medical community as a new antibiotic isolated in Spain. As well as using the article to promote CEPA, in whose laboratories the antibiotic had actually been obtained, Gallego took the opportunity to highlight the need to promote research in Spain, both through public policies and private initiatives. It was investment by CEPA and Merck, Gallego stressed, that had made the new antibiotic possible, and it was this aspect that dominated reports in the daily newspapers: 'Spanish industry does research'.[56] Gallego was cautious regarding the antibiotic capacity of the new drug, as the therapeutic properties of phosphonomycin had yet to be validated.[57]

Some days later, a Madrid daily newspaper declared phosphonomycin the result of '15 years of research, 12 thousand samples, and one million two hundred actinomycetes cultures analysed'.[58] Merck and CEPA had collaborated in an investigative enterprise that had not received a penny of public money. These publications presented the particular case of an antibiotic isolated in Spain, in what the participating scientists considered to be a Spanish endeavour carried out with Merck technical support. This was very similar to Waksman and Schatzs's discovery of streptomycin, the identification of which relied on the chemical skills at Merck Research Laboratories.

The movement of agents—scientists, engineers, methods and materials—between Madrid and Rahway illustrates the circulation of practices and knowledge in the history of a new antibiotic presented to the Spanish public as a national achievement with a date and place of birth. By 1969, Spanish autarchy as a central policy of Franco's government had been replaced by a growing number of technology transfers through contracts with foreign companies, whose methods of production were used by the entire industrial sector.[59] This situation made the production of the first Spanish antibiotic a highly unusual event; at least this was the message presented to the newspapers' readers.

Both the product and the preparation process were eligible for patent protection, and this was granted at the US Patent Office on February 1, 1972. The patent for 'Antibiotic 833A' was signed by David Handlin and Edward O. Stapley, from Merck, and Sagrario Mochales and Justo Martínez Mata, from CEPA. The procedure for synthesis of the new antibiotic was also patented in Spain. Mochales remembers that her name had not originally been included—neither women nor technicians usually signed—but Heindlin insisted, and her name can be found on both the application and the 1972 patent awarded by the US Patent Office.[60] By the time of the application, Sagrario Mochales was already married and pregnant, and as the director did not consider it suitable that a married woman keep a job outside the home, her job was at risk. According to her recollections, Mochales's signature on the patent application protected her: a woman with a patent could not be fired. With support from Carmela Viejo she was allowed to stay. Although tensions remained for some time, eventually the CEPA director accepted her as a married woman, a researcher and a mother.

Fosfomycin in the Clinic and the Factory

Phosphonomycin, later named fosfomycin according to the US Adopted Names Council, was tested against a series of antibiotic-resistant cultures.[61] As still an experimental medical drug, the research procedure included animal and clinical testing, as part of a linear scheme leading from the laboratory to the clinic and industrial production line.[62] The sodium salt produced 'reasonable' activity when tested in mice and a group of healthy volunteers: 'male prisoners after they had been informed of the nature of the agent and of the studies to be performed'. It showed an 'apparent lack' of toxicity and was easily eliminated in urine, which suggested that clinical trials, especially those of the urinary tract, were 'warranted'.[63] Similar results were achieved with the calcium salt.[64]

Preliminary clinical trials in six female and fourteen male patients at the Wilmington Medical Center, Delaware, demonstrated 'satisfactory' results in a small number of urinary tract infections, while in others it appeared less promising. However, six patients with infections resistant to other antibiotics were successfully treated, with three of these requiring 'maintenance therapy'. In patients displaying resistance to previously used antibiotics, phosphonomycin had limited success, and some patients developed resistances, as did some in vitro strains.[65] Not recommended in gonococcal infections, it appeared active against staphylococcus infections, for which penicillin was already being used.[66] *Klebsiella* infection remained resistant to phosphonomycin.[67] As there is no relevant archival material as yet, it is difficult to analyse the results of these trials in depth: personal recollections and academic papers provide rich information, but are of limited value without any other preserved testimony.

As fermentation was not an efficient method, Merck decided to establish a synthetic procedure on a pilot plant scale. Both oral and injectable presentations of the drug were required. The pilot plant enabled a large enough amount of the new antibiotic to be obtained for clinical and toxicological tests. A standardised protocol composed of a set of tests was followed to assure the efficacy and non-toxicity of the new product.[68] The synthesis process had the usual difficulties in separation of the enantiomers (stereoisomers, the separation of which took place at the final step of synthesis in salt form). By the end of 1969 the pilot plant's report was ready.

As recalled by the two woman scientists involved in the early detection and testing of the sample containing the new antibiotic, Sagrario Mochales and Isabel Martín, the substance's activity was high when administered intramuscularly, but lower when taken orally.[69] According to Carlos Rosas, the chemical engineer who developed the pilot plant project, Merck decided not to take the project any further based on an unfavourable sales forecast.[70] By the late 1960s and early 1970s, oral administration was increasingly being considered best practice for most drugs, despite awareness that intravenous and later intramuscular injection produced a swifter effect. Consumption of pharmaceuticals was increasing rapidly and oral intake made this consumption easier: injection was regarded a practice for medical staff, although many women had become skilled at injecting their family members. By that time, a preference for oral intake had become one of the features of Margaret Sanger's long quest to find researchers able to provide women with a contraceptive drug.[71]

A Statement before Congress supposedly made by US Surgeon General William H. Stewart in 1969 also discouraged investment in antibiotic production: 'the time has come to close the book on infectious diseases'. Although this Statement has been deemed an 'urban legend', by the late 1960s and 1970s general opinion held that the era of infectious disease was coming to an end, and pharmaceutical firms began to close down their antibiotic screening programmes.[72] This environment, together with the low activity when administered orally, could have influenced Merck's decision to leave industrial production and marketing of the new antibiotic to CEPA.

Apparently dismissing the oral consumption issue, CEPA, which had rights to sell the drug in Spain, Portugal and Morocco, decided to carry on and made the project a priority.[73] The Merck Department of Chemical Engineering provided technical support with construction of the factory, started in 1972. The requirements tested in the pilot plant for chemical synthesis were put into practice, a change from the manufacturing methods previously used in CEPA antibiotic production. Assembly of the factory, according to Merck engineer Carlos Rosas, was skilful and economical. Equipment in the new building included reactors, large steel tanks of between 1000 and 5000 litres—among them tank wagons abandoned by the Spanish railway company and recycled by one of the supervisors—an automatic ultracentrifuge, a rotary vacuum dryer and some equipment to retrieve and reuse one of the intermediate chemicals.[74] The building of a factory for chemical production of an antibiotic was a new challenge for CEPA, a firm that had previously specialised in fermentation procedures.[75]

Scientific Research and Marketing

Under its final name, *fosfocina*, the drug was marketed as a 'new broad-spectrum antibiotic', acidic, water soluble, electrophoretically mobile, cell-wall active (by inhibiting cell-wall synthesis) with unique antibacterial and cross-resistance spectra, clearly distinguishing it from all antibiotics previously studied in the programme.[76] It was manufactured in Aranjuez from 1973, with Merck keeping the right to market it in the rest of Europe by paying four per cent of sales to CEPA. The Spanish firm rolled out an advertising campaign when *fosfocina* went on sale in March 1973. Marketed for oral, intramuscular and intravenous administration, four pages of advertisements presenting the drug in capsules, syrup and vials for injections were published in the weekly medical magazine, *Noticias Médicas* (see Fig. 6.3).

A symposium on fosfomycin funded by CEPA was held in Madrid in July 1975. As a scientific celebration of the research programme's success, the conference appears to have been a *mise-en-scène* of the therapeutic activity of a small new molecule of low molecular weight and a new chemical structure; for an antibiotic whose antimicrobial activity had been tested in clinical and research laboratories all around Spain and abroad, and at Merck Research Laboratories. It brought fosfomycin back to its original landscape, the research laboratory, to the microbiological methods which had made it visible, detectable in Petri dishes and active in laboratory animals and medical practice.[77]

The Research Tool Revisited: Fosfomycin's Mechanism of Action

Fosfomycin was the first cell-wall active agent to be found through a focus on inhibition of the bacterial cell wall. CEPA and Merck researchers kept the new antibiotic within a research project that used fosfomycin both as a tool and as a rationale for investigating the mode of its

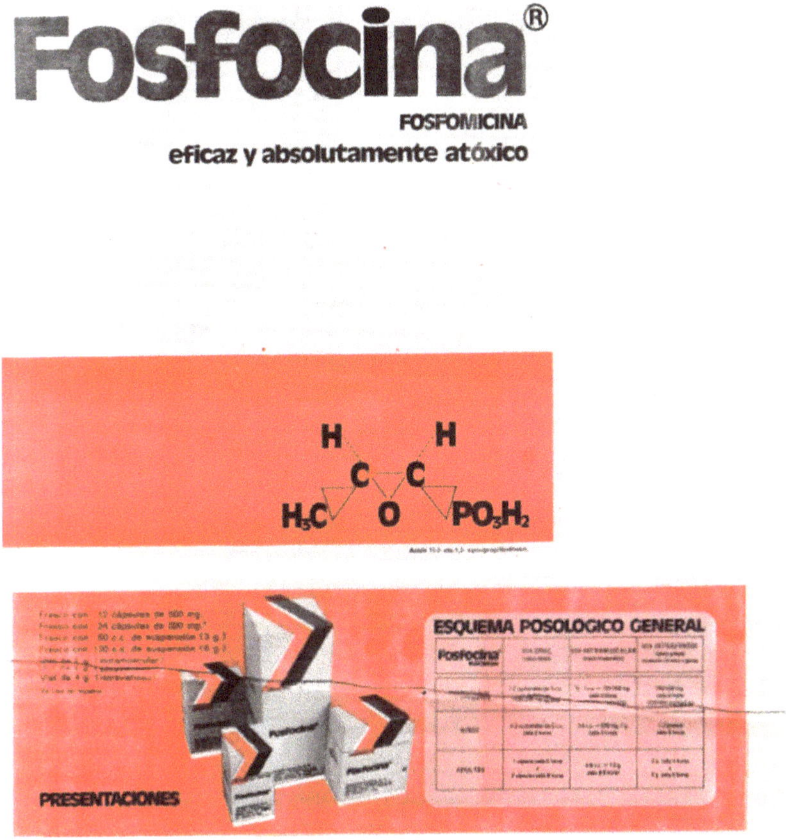

Fig. 6.3 *Fosfocina* advertisements published in *Noticias Médicas*, March 1973

biochemical action. This research agenda contributed to biochemical knowledge of the cell wall while maintaining the antibiotic as a biochemical research subject in its own right. As explained above, Merck researchers were not the only—or the first—experimenters to use an antibiotic in biochemical research. The mode of action of other antibiotics, namely penicillin, was already known. As a research endeavour, initial determination of penicillin's action can be attributed to Park, Lederberg and Strominger between 1956 and 1957, as has been narrated above; later research would focus on the action of enzymes in different phases.

Merck research activities took part in the biochemical developments of the time by describing fosfomycin as an inactivating agent of the first step in the biosynthesis of the cell wall, the one in which P-enolpyruvate is involved. It turned out that the enzyme that participates in the reaction of this remarkable biochemical substance, phosphoenolpyruvate, was a transferase, which transfers the enolpyruvil to a longer, more complex molecule, as it 'is irreversibly inhibited by the antibiotic fosfomycin'. The similarities between fosfomycin and the intermediate of such a transfer were identified by Merck researchers Patrick Cassidy, Jean and Frederick Kahan, and Helmut Kropp.[78] Cassidy, and Frederick and Jane Kahan suggested a mechanism for the direct chemical participation of fosfomycin in the inactivation of an enzyme involved in the synthesis of the cell wall, in which the stereochemical and composition features of fosfomycin fit beautifully.

Due to the structure and composition similarities between fosfomycin and pyruvate, the inactivation of the enzyme in vitro produced the bonding of fosfomycin to the protein component of the cell wall. Cell-wall synthesis was inhibited by this mechanism, according to the authors, who continued working on this project by connecting microbiology with biochemistry, thus contributing to the new disciplinary space cell biology was becoming at that time.[79] With this research project, Merck laboratories and fosfomycin joined in the research agenda of cell biology, while also keeping antibiotics among its instrumental tools in the research laboratory.

Geographies of Fulfilled Promises: The Factory and the Research Laboratory

From the end of the Second World War, the future of the West was being invented. One of the forms this future took was in new substances used to cure infection. The cure was the future, one free from infection, and the materials of this future were the drugs themselves. Recovery from the Second World War was the agent that made this cure available through the distribution of penicillin. Antibiotics had been among the techniques and commodities that brought the war to an end, and now they were received as agents for the future, as technological and scientific progress, and a promise of prosperity based on the research laboratory and expert knowledge. The promises to be fulfilled—of making every infection disappear—had been created by penicillin during a long, tragic world war. The war was won, or at least brought to a close, by technological achievements besides the atomic bomb and radar. Among the Second World War technologies, penicillin stood out: a new drug that healed infected wounds was seen as an unprecedented marvel. In this environment of needs and promises based on science and technology, *fosfocina* fulfilled the promise of discovery embodied in the screening research programme established at CEPA in collaboration with Merck, based in microbiology.

The bacterial cell wall proved to be the target for penicillin. This targeting made the cell produce protoplasts. The cell would lose its wall but not its membrane, and this loss made the microbe more permeable to substances that would halt its growth.[80] The practices for obtaining this particular cell part were transformed into a method by Merck researcher Eugene Delaney to detect antimicrobial activity in soil samples of unknown composition. A confirmation test carried out with a chromogenic substance enabled CEPA researcher Sagrario Mochales to detect a new antimicrobial activity in 1966. A lyophilised concentrate of the culture was sent to the Merck Research Laboratories in Rahway, where the new antibiotic was isolated and characterised: a three-bonding ring of a phosphonic acid.

The screening programme was part of the biography of CEPA, a firm whose main task was to manufacture antibiotics, and whose research was conceived as support for this manufacturing.[81] Created to produce penicillin in Spain, CEPA became an industrial firm that also carried out research, with a small department established inside the bottling plant. Appropriation of the drug took place after a long process: almost 20 years passed between the outset of the screening programme and marketing of the new antibiotic. Having closely followed the hierarchical structure imposed by Merck in the agreements that created the programme, success originated from the microbiological knowledge and skills of a young woman researcher who had been with the project since its inception in 1954, and shared the ambitions of the pharmaceutical firms that hired her. This was the fulfilment of a collective ambition, shared by everyone working on the industrial research project, from the director of the programme at CEPA, Antonio Gallego, to Mochales, a scientist working in a research team.

Merck researchers kept the new antibiotic within a research project that used fosfomycin both as a tool and as a rationale for investigating the mode of biochemical action of a new antimicrobial. Bacterial wall synthesis was inhibited by a biochemical mechanism produced by the antibiotic, according to the Merck scientists who deciphered it.[82] With this research project, CEPA and Merck laboratories, and fosfomycin, joined in the research agenda of cell biology, as they carried out research on the bacterial wall using the antibiotic as a tool. By constituting a network of disciplinary and professional spaces related to both biological and medical practices, cell biology was part of a growing field that was starting to be called biomedicine.[83]

Fosfomycin created higher expectations: new antibiotics would be found if Petri dishes in which samples were cultured were examined carefully enough. The title given to fosfomycin by Merck—MK-0955—was kept and the drug itself would remain a Merck achievement, despite the US firm abandoning production plans.[84] The Spanish firm maintained its industrial objective and with the assistance of Merck engineers built a factory for chemical synthesis of the new antibiotic in Aranjuez, near Madrid. A new space appeared to have been opened up for Spanish pharmaceutical products, both at home and abroad. Although there are no sale figures for fosfomycin in CEPA annual reports, exports of both in-bulk and bottled batches are mentioned, as are plans to license fosfomycin in Latin America and Japan. Commercial success thus followed scientific and medical accomplishments.[85] Fosfomycin shared manufacturing space in CEPA factories with new drugs, such as amplicillin—a successful, semisynthetic penicillin—with a patent from the Japanese Yamanouchi Pharmaceutical Co., and erythromycin from the Italian firm, Prodotti-Terapeuticci.

The new drug proved to be not only a future product of its near past, but a commodity originating from the early promotion of basic research, spurring industrial innovations to fulfil the promise of the research laboratory in the antibiotic era. The travels of knowledge and practices, of policies and industrial manufacturing procedures in peacetime, participated in the recovery of the East, West, North and South of Europe, the North Americas and Asia. As a foreign medical drug, the effects of penicillin went beyond healing to become a promise of progress.[86] A project by a professor of physiology of the University of Madrid Medical School, Antonio Gallego, who placed his belief in research and the industrial manufacture of antibiotics in Spain, became a new antimicrobial drug, identified by the young researcher, Sagrario Mochales, in the CEPA–Merck collaborative programme. Mochales later worked on extracts from the drug with Isabel Martín, who had joined the group by the time fosfomycin was detected and identified. While antibiotics stabilised their agency in the laboratory of cell biology, contributing to knowledge about cellular processes through their use as research tools, some—penicillin, fosfomycin and others—exhibited a long life. Or perhaps, as Toine Pieters and Stephen Snelders have suggested in relation to psychotropic drugs, antibiotics exhibit career cycles, fluctuations in consumption, exemplified in the reappearance of fosfomycin during the 1990s, re-presented as a treatment for urinary infections still being prescribed today.[87]

Notes

1. 'Curar la infección urinaria en un día', *La Vanguardia*, February 15, 1991, Suplemento Medicina y Calidad de Vida, 9. Strikingly, the article ended by noting that clinicians had long awaited approval for the marketing of fosfomycin. Apparently, it was the presentation as a water-soluble suspension that was new.
2. The annual reports of CEPA board meetings are preserved at Archivos de CEPA, Fondos Urquijo, Instituto Carlos III-Juan March de Ciencias Sociales, Universidad Carlos III de Madrid, Getafe, Madrid.
3. On the lives of drugs and their trajectories as analytical tools, see Jean-Paul Gaudilliére and Volker Hess, eds., *Ways of Regulating Drugs in the 19th and 20th Centuries* (London: Palgrave Macmillan, 2012). On the proposal of 'career cycles' of drugs, relating to their use for purposes other than their early indications, see Stephen Snelders, Charles Kaplan and Toine Pieters, 'On cannabis, chloral hydrate, and career cycles of psychotrophic drugs in medicine', *Bulletin of the History of Medicine* 80 (2006): 95–114.
4. On post-war biomedicine shaped by a focus on cells and molecules, see Jean-Paul Gaudillière, *Inventer la biomédecine: la France, l'Amérique et la production des savoirs du vivant, 1945–1965* (Paris: La découverte, 2002). In part inspired by Gaudillière's approach, Keating and Cambrosio have suggested that biology has encircled medicine: Peter Keating and Albert Cambrosio, *Biomedical Platforms: Realigning the Normal and the Pathological in Late-Twentieth-Century Medicine* (Cambridge, MA: MIT Press, 2003).
5. Scott H. Podolsky, *The Antibiotic Era: Reform, Resistance, and the Pursuit of a Rational Therapeutics* (Baltimore: Johns Hopkins University Press, 2014). For a history of resistances *avant la lettre*, see Christoph Gradmann, 'Magic bullets and moving targets: Antibiotic resistance and experimental chemotherapy, 1900–1940', *Dynamis* 31 (2011): 29–46.

6. Lynn L. Silver, 'Rational approaches to antibacterial discovery: Pre-genomic directed and phenotypic screening', in *Antibiotic Discovery and Development*, eds. T. J. Dougherty and M. J. Pucci (New York: Springer, 2012), 33–75; and R. H. Baltz, 'Antibiotic discovery from actinomycetes: Will a renaissance follow the decline and fall', *SIM News* 55 (2005): 186–96.
7. Eugene L. Dulaney, A. H. Larsen and Edward O. Stapley, 'A note on the isolation of microorganisms from natural sources', *Mycologia* 47 (1955): 420–2.
8. Edward O. Stapley, 'Cross-resistance studies and antibiotic identification', *Applied Microbiology* 6 (1958): 392–8.
9. Selman Waksman, 'Streptomycin: Background, isolation, properties, and utilization', *Nobel Lecture*, December 12, (1952): 370–88.
10. H. Boyd Woodruff, Sebastián Hernández and Edward O. Stapley, 'Evolution of an antibiotic screening programme. A tribute to Justo Martínez Mata', *Hindustan Antibiotics Bulletin* 21 (1979): 71–84.
11. Silver, 'Rational approaches to antibacterial discovery'; Hans H. Gadebusch, Edward O. Stapley and Sheldon B. Zimmerman, 'The discovery of cell wall active antibacterial antibiotics', *Critical Reviews in Biotechnology* 12 (1992): 225–43.
12. On the history of drug patenting see Jean-Paul Gaudillière, 'Professional or industrial order? Patents, biological drugs, and pharmaceutical capitalism in early twentieth century Germany', *History and Technology* 24 (2008): 107–33; Ana Romero de Pablos, 'Regulation and the circulation of knowledge: Penicillin patents in Spain', *Dynamis* 31 (2011): 87–107; Ana Romero de Pablos, 'Patents, antibiotics, and autarky in Spain', *Medicina nei Secoli* 26 (2014): 423–49; Axel C. Hüntelmann, 'Priority, property and trust: Patent laws and pharmaceuticals in the German Empire', *Interdisciplines* 2 (2012): 194–226.
13. E. O. Stapley, J. M. Mata, I. M. Miller, T. C. Demny and H. B. Woodruff, 'Antibiotic MSD-235. I. Production by streptomices avidinii and streptomyces lavandulae', *Antimicrobial Agents and Chemotherapy* 3 (1963): 20–7.
14. On the use of antibiotics in the same period by British biochemist Ernest Gale at Cambridge, see Hans-Jörg Rheinberger, 'Comparing experimental systems: Protein synthesis in microbes and in animal tissue at Cambridge (Ernest F. Gale) and at the Massachusetts General Hospital (Paul C. Zamecnik), 1945–1960', *Journal of the History of Biology* 29 (1996): 387–416.
15. Silver, 'Rational approaches to antibacterial discovery'.
16. Scott H. Podolsky, 'Antibiotics and the social history of the controlled clinical trial', *Journal for the History of Medicine and Allied Sciences*

65 (2010): 327–67; Podolsky, *The Antibiotic Era*. Microbiology took part in the construction of resistance testing; see Gradmann, 'Magic bullets and moving targets'.

17. For an inspiring conceptual approach to the bedside and the research laboratory, see Ilana Löwy, *Between Bench and Bedside: Science, Healing, and Interleukin-2 in a Cancer Ward* (Cambridge, MA: Harvard University Press, 1996).
18. A. D. Gardner, 'Morphological effects of penicillin on bacteria', *Nature* 146 (1940): 837.
19. L. D. Smith and Telma Hay, 'The effect of penicillin on the growth and morphology of *Staphylococcus aureus*', *J. Franklin Inst.* 233 (1942): 598–602, 602.
20. On Gale, see Rheinberger, 'Comparing experimental systems'.
21. Jean-Marie Ghuysen, 'The concept of the penicillin target from 1965 until today: The thirteenth Marjory Stephenson Memorial Lecture', *Journal of General Microbiology* 101 (1977): 13–33, 13. On Marjory Stephenson, see Sonia Štrbáňová, *Holding Hands with Bacteria: The Life and Work of Marjory Stephenson* (Berlin: Springer, 2016).
22. Milton R. J. Salton, 'Cell structure and the enzymic lysis of bacteria', *Journal of General Microbiology* 9 (1953): 512–23.
23. Milton R. J. Salton, 'Cell wall of *Micrococcus lysodeikticus* as the substrate of lysozyme', *Nature* 170 (1952): 746; Salton, 'Cell structure', 513.
24. Salton, 'Cell structure', 521.
25. On the cell membrane in later years, see Mathias Grote and Maureen A. O'Malley, 'Enlightening the life sciences: The history of halobacterial and microbial rhodopsin research', *FEMS Microbiology Reviews* 35 (2011): 1082–99; Mathias Grote, 'Surfaces of action: Cells and membranes in electrochemistry and the life sciences', *Studies in History and Philosophy of Biological and Biomedical Sciences* 41 (2010): 183–93; Mathias Grote, 'Purple matter, membranes and "molecular pumps" in rhodopsin research (1960s–1980s)', *Journal of the History of Biology* 46 (2013): 331–68.
26. H. H. Martin, 'Bacterial protoplasts—A review', *Journal of Theoretical Biology* 5 (1963): 1–34, 1; See also Joshua Lederberg and Jacqueline St Clair, 'Protoplasts and L-type growth of *Escherichia coli*', *Journal of Bacteriology* 75 (1958): 143–60.
27. Joshua Lederberg, 'Bacterial protoplasts induced by penicillin', *Proceedings of the National Academy of Sciences of the United States of America* 42 (1956): 574–8.
28. Joshua Lederberg, 1958, Nobel lecture: A view of genetics. *Nobelprize.org*. Nobel Media at http://www.nobelprize.org/nobel_prizes/medicine/laureates/1958/lederberg-lecture.html, 10.

29. Angela N.Creager, 'Adaptation or selection? Old issues and new stakes in the postwar debates over bacterial drug resistance', *Studies in History and Philosophy of Biological and Biomedical Sciences* 38 (2007): 159–90; Gradmann, 'Magic bullets'; Anne Kveim Lie, 'Producing standards, producing the Nordic region: Antibiotic susceptibility testing, from 1950–1970', *Science in Context* 27 (2014): 215–48, at https://doi.org/10.1017/S0269889714000052.
30. Lederberg and St Clair, 'Protoplasts and L-type growth', 157.
31. See Joshua Lederberg's correspondence during these years in Joshua Lederberg Papers, Series Correspondence, 1935–2002; *SubSeries*: 1953–1960; US National Library of Medicine, at http://profiles.nlm.nih.gov/ps/retrieve/Series/3 (hereafter Lederberg Papers), accessed March 15, 2017.
32. Joshua Lederberg, 'Mechanism of action of penicillin', *Journal of Bacteriology* 73 (1957): 144; James T. Park and Jack L. Strominger, 'Mode of action of penicillin', *Science* 125 (1957): 99–101.
33. Park to Lederberg, October 8, 1956; Strominger to Lederberg, March 5, 1957; Lederberg Papers. Accessed January 15, 2015.
34. See M. J. Santesmases and E. Suárez-Díaz, 'A cell-based epistemology: Human genetics in the era of biomedicine', *Historical Studies in the Natural Sciences* 45 (2015): 1–13.
35. This discussion on terms and meanings relating to the bacterial cell wall evokes that concerning the protoplasmic theory of life in the late nineteenth century. See Gerald L. Geison, 'The protoplasmic theory of life and the vitalist–mechanist debate', *Isis* 60 (1969): 273–92. On its connection with the later theory of enzymes at the origins of biochemistry, see Robert E. Kohler, 'The enzyme theory and the origin of biochemistry', *Isis* 64 (1973): 181–96. This historiography of previous accounts has informed biology as a historicised epistemological space.
36. Kenneth McQuillen and Milton R. J. Salton, 'Synthetic media for maintenance and induction of lysogenic *Bacillus megaterium*', *Biochimica et Biophysica Acta* 16 (1955): 596–7; Milton R. J. Salton and Kenneth McQuillen, 'Bacterial protoplasts: II. Bacteriophage multiplication in protoplasts of sensitive and lysogenic strains strains of *Bacillus megaterium*', *Biochimica et Biophysica Acta* 17 (1955): 465–72.
37. Kenneth McQuillen, 'Bacterial protoplasts: growth and division of protoplasts of *Bacillus megaterium*', *Biochimica et Biophysica Acta* 18 (1955): 458–61.
38. 'Proposed note to Nature. Concerning the use of the term protoplast', undated, McQuillen to Lederberg, filed as of 1957. Capitals in the original. Joshua Lederberg Papers. Accessed March 15, 2017.

39. McQuillen to Lederberg, November 24, 1956; Lederberg to McQuillen, December 7, 1957. Joshua Lederberg Papers. Accessed March 15, 2017.
40. McQuillen to Lederberg, December 14, 1957. Joshua Lederberg Papers. Accessed March 15, 2017.
41. S. Brenner, F. A. Dark, P. Gerhardt, M. H. Jeynes, O. Kandler, E. Kellenberg, E. Klieneberger-Nobel, K. McQuillen, M. Rubio-Huertos, M. R. J. Salton, R. E. Strange, J. Tomsik and C. Weibull, 'Bacterial protoplasts', *Nature* 181 (1958): 1713–15.
42. Kenneth McQuillen, 'Bacterial protoplasts', in *The Bacteria: A Treatise on Structure and Function*, eds. I. C. Gunsalus and R. Y. Stanier (New York and London: Academic Press, 1960), 249–359, on 257–8.
43. Jacques Coyette, Jean-Marie Frère and Peter Reynolds, 'Jean-Marie Ghuysen', *Molecular Microbiology* 57 (2005): 871–3.
44. Wolfhard Weidel and H. Pelzer, 'Bag-shaped macromolecules. A new outlook on bacterial cell walls', *Advances in Enzymology* 26 (1964): 193–232.
45. See Silver, 'Rational approaches to antibacterial discovery', and references therein.
46. I owe these descriptions of the Perm and Sphero methods to Isabel Martín and Sagrario Mochales. A later, geneticised version of this test appears in Silver, 'Rational approaches to antibacterial discovery'.
47. E. O. Stapley, D. Hendlin, J. M. Mata, M. Jackson, H. Wallick, S. Hernández, S. Mochales, S. A. Currie and R. M. Miller, 'Phosphonomycin I. Discovery and in vitro biological characterization', *Antimicrobial Agents and Chemotherapy* 9 (1968): 284–90, 286.
48. Woodruff, Hernández and Stapley, 'Evolution of an antibiotic screening program', 73; J. Martínez Mata, 'La investigación de antibióticos en la industria farmacéutica española. Historia de un descubrimiento: la fosfomicina', *Química e Industria* 20 (1974): 51–9. On the procedure, see K. Miller, 'Antibiotic MSD-235. III Evaluation of antibiotic MSD-235 in vitro and in experimental infections in mice', *Antimicrobial Agents and Chemotherapy* (1969): 33–7.
49. D. Hendlin, E. O. Stapley, M. Jackson, H. Wallik, A. K. Miller, F. J. Wolf, T. W. Miller, L. Chaict, F. M. Kahan, E. L. Foltz, H. B. Woodruff, J. M. Mata, S. Hernández and S. Mochales, 'Phosphonomycin: A new antibiotic produced by strains of streptomyces', *Science* 166 (1969): 122–3; followed by B. G. Christinesen, W. J. Leanza, T. R. Beattie, A. A. Patchett, B. H. Arison, Robert E. Ormond, Frederick A. Kuehl, Jr., G. Albers-Schonberg and O. Jardetzky, 'Phosphonomycin: Structure and synthesis', *Science* 160 (1969): 123–5.
50. Woodruff, Hernández and Stapley, 'Evolution of an antibiotic screening program'.

51. The Interscience Conference on Antimicrobial Agents and Chemotherapy (ICAAC) was the American Society for Microbiology's annual meeting on infectious disease.
52. The collection at the NRRL was created in 1940. The naming system is not clear. On the history of the US Agricultural Research Service Culture Collection, see http://nrrl.ncaur.usda.gov/TheCollection/.
53. Hendlin et al., 'Phosphonomycin'; Christinesen et al., 'Phosphonomycin'.
54. The order of signatories on an academic, scientific paper denotes hierarchies. They change over time and can differ between disciplines, but always reflect authority management.
55. *Noticias Médicas*, October 15, 1969.
56. 'La Industria española investiga', *La Vanguardia*, October 3, 1969, 27. The report, signed by the Spanish press agency EFE in Washington, refers to the meeting of the American Society for Microbiology.
57. 'Fosfomicina, nuevo antibiótico de amplio radio de acción', *La Vanguardia*, October 31, 1969, 74.
58. *Pueblo*, November 25, 1969.
59. See Mar Cebrián, 'Technological Imitation and Economic Growth during the Golden Age in Spain: 1959–1973' (PhD diss., Florence: European University Institute, Florence, 2004). Also Santiago López and Mar Cebrián, 'Economic Growth, Technology Transfer, and Convergence in Spain, 1960–1973', in *Technology and Human Capital in Historical Perspective*, eds. J. Ljunberg and J.-P. Smits (New York: Palgrave-Macmillan, 2004), 120–44.
60. D. Hendlin, E. O. Stapley, S. Mochales and J. Martínez Mata (1972), assignors to Merck & Co., Inc., Rahway, NJ. Antibacterial composition containing (-)(cis-1,2-epoxipropyl) phosphoric acid, United States Patent Office. Patented February 1, 1972.
61. A, Gallego, A. Rodríguez and J. Martínez Mata, 'Fosfomycin: Pharmacological studies', *Drugs Today* 10 Supplement (1974): 161–8; Stapley et al., 'Phosphonomycin I'.
62. I found no archival material on the trials, but did find three collections of papers on the new drug published in 1969, 1970 and 1977. The 1969 collection was the result of the earliest test and trials in the USA and was published in the 1969 volume of the by then well-known collection—not yet a journal—edited by penicillin expert Gladys Hobby, *Antimicrobial Agents and Chemotherapy.* On the history of the journal, which shifted from an annual book series in the 1950s to a journal in the mid-1960s, see G. A. Jacoby, 'History of antimicrobial agents and chemotherapy from 1972 to 1998', *Antimicrobial Agents and Chemotherapy* 43 (1999): 999–1002. On early medical trials in the USA, see H. M. Marks, *The Progress of Experiment: Science and Therapeutic Reform in*

the United States, 1900–1990 (Cambridge: Cambridge University Press, 2000). The collection from 1970 was published in *Anales del Instituto de Farmacología Española*, volume 19 (1970), on bacterial sensitivity to the new drug by the members of CEPA research group: Isabel Martín, Sagrario Mochales, Sebastián Hernández, A. Rodríguez and Justo Martínes Mata. The third collection was published in a special supplement of the Swiss-based journal *Chemotherapy* 23 (1977): suppl. 1.

63. E. L. Foltz and H. Wallick, 'Pharmacodynamics of phosphonomycin after intravenous administration in man', *Antimicrobial Agents and Chemotherapy* 9 (1969): 316–21.
64. E. L. Foltz, H. Wallick and C. Rosenblum, 'Pharmacodynamics of phosphonomycin after oral administration in man', *Antimicrobial Agents and Chemotherapy* 9 (1969): 322–6.
65. H. Clark, N. K. Brown, J. F. Wallace and M. Turck, 'Evaluation of phosphomycin, a new cell wall-active antibiotic', *Antimicrobial Agents and Chemotherapy* 9 (1969): 338–42.
66. J. W. Smith and J. P. Sanford, 'Activity of phosphonomycin in nasal carriers of coagulase positive Staphylococci', *Antimicrobial Agents and Chemotherapy* 9 (1969): 346–8.
67. Jack A. Barnett, Paul M. Southern Jr, James P. Luby and Jay P. Sanford, 'Eficacy of phosphonomycin in treatment of urinary-tract infections', *Antimicrobial Agents and Chemotherapy* 9 (1969): 349–51.
68. Carlos Rosas, retired Merck chemical engineer, to the author, November 11, 2002.
69. On its low cure rate when administered orally, see W. Holloway, J. Clark and R. Rocco, 'Preliminary clinical trials with phosphomycin', *Antimicrobial Agents and Chemotherapy* 9 (1969): 327–31.
70. Carlos Rosas, to the author, November 11, 2002.
71. Lara Marks, *Sexual Chemistry: A History of the Contraceptive Pill* (New Haven: Yale University Press, 2010); Nelly Oudshoorn, *Beyond the Natural Body: An Archaeology of Sex Hormones* (London: Routledge, 2003); Elisabeth S. Watkins, *On the Pill: A Social History of Oral Contraceptives, 1950–1970* (Baltimore: Johns Hopkins University Press, 2011).
72. B. Spellberg, 'Dr. W. H. Stewart, mistaken or maligned?' *Clinical Infectious Disease* 47 (2008): 294.
73. Gallego to Juan Lladó, November 14, 1974. Antonio Gallego, personal archive.
74. Carlos Rosas, to the author, November 11, 2002.
75. Compañía Española de Penicilinas y Antibióticos, 'Memoria del Ejercicio 1972'. Archives of Servicio de Estudios del Banco Urquijo. Microfiches preserved at the Biblioteca de Ciencias Sociales, Fundación Juan March, Madrid.

76. Stapley et al., 'Phosphonomycin I'. On the concept of scientific marketing, see Jean-Paul Gaudillière and Ulrike Thoms, 'Pharmaceutical firms and the construction of drug markets: From branding to scientific marketing', *History and Technology* 29 (2013): 105–15.
77. A special supplement of the Swiss-based journal *Chemotherapy*—previously *Chemoterapia*—published a collection of papers presented at the symposium. See the introduction, A. Gallego and J. M. Rubio, eds., '1977 Fosfomycin. Proceedings of the international symposium on fosfomycin, Madrid, 10–11 July 1975', *Chemotherapy* 23 (1977): suppl. 1.
78. Patrick J. Cassidy and Frederick M. Kahan, 'A stable enzyme phosphoenolpyruvate intermediate in the synthesis of Uridine-5'-diphospho-N-actey l-2-amino-2-deoxyglucose 3-0-enolpyruvyl ether', *Biochemistry* 12 (1973): 1364–75.
79. Frederick M. Kahan, Jean S. Kahan, Patrick J. Cassidy and Helmut Kropp, 'The mechanism of action of fosfomycin (Phosphonomycin)', *Annals of the New York Academy of Sciences* 235 (1974): 364–87.
80. On membranes as places of biochemical and cultural action, see Grote, 'Surfaces of action' and Grote, 'Purple matter'.
81. Gadebusch, Stapley and Zimmerman, 'The discovery of cell wall active antibacterial antibiotics'. Merck was a more diversified company. It was already in the 1950s described by the public relations executive and author Tom Mahoney as a leader in vitamins and nutritional supplements, penicillin and streptomycin, steroids, and also in vaccines and sera manufacturing. Tom Mahoney. *The Merchants of Life: An Account of the American Pharmaceutical Industry* (New York: Harper, 1959). On vaccines at Merck, see Louis Galambos and Jane Eliot Sewell, *Networks of Innovation: Vaccine Development at Merck, Sharp and Dohme, and Mulford, 1895–1995* (Cambridge: Cambridge University Press, 1997).
82. Kahan et al., 'The mechanism of action of fosfomycin'.
83. On biomedicine as a new term see S. de Chadarevian and H. Kamminga, 'Introduction', in *Molecularizing Biology and Medicine: New Practices and Alliances, 1910s–1970s*, eds. S. de Chadarevian and H. Kamminga (Amsterdam: Harwood, 1998).
84. See R. A. Fromtling, 'Fosfomycin (MK0955): An overview', *Drugs of Today* 23 (1987): 151–8.
85. Gallego-Lladó correspondence 1971. Antonio Gallego, personal archive.
86. If I may reconfigure the paired term, fear and progress, which historian Antonio Cazorla has argued was a feature of life under the early Franco dictatorship. See Antonio Cazorla Sánchez, *Fear and Progress: Ordinary Lives in Franco's Spain, 1939–1975* (Oxford: John Wiley and Sons, 2009).
87. Snelders, Kaplan and Pieters, 'On cannabis'.

CHAPTER 7

Beyond Healing: Antibiotic Resistance and Regulatory Regimes as Agents in the Spanish Transition to Democracy

Legislation relating to antibiotic prescription in Spain will be reconstructed here as a history of regulation and consumption control. Such control was largely political in the early days of penicillin's circulation (Chaps. 1 and 2), embedded in the bureaucracy of Franco's regime. The first decade of penicillin in Spain was one of scarcity and a strict control of sales; consumption remained uncontrolled due to illegal circulation. Domestically produced penicillin was available without prescription from the early 1950s, but concern about resistance though the 1950s and 1960s prompted the gradual introduction of regulations and prescription requirements. Clinicians, in Spain and abroad, had expected and witnessed cures by penicillin since its early days, but as cases where predicted healing did not occur began to appear, awareness of resistance increased, particularly in the research laboratory. The old miracle drug had enjoyed such longevity it was almost as if penicillin had always existed, shielding humanity from the fear of death through infection. Still manufactured and marketed today, though in different presentations, the beta-lactam ring featuring its particular chemical structure has been a sign of hope among marketed antimicrobial molecules for decades.

The political scenario during the dictatorship endured, with no negotiation regarding decisions on the regulation, consumption and industrial production of commodities while the public renown of antibiotics overshadowed the issue of resistance. The end of Spain's isolation during the 1940s, the promising 1950s and particularly the subsequent economic

M. J. Santesmases, *The Circulation of Penicillin in Spain*, Medicine and Biomedical Sciences in Modern History,
https://doi.org/10.1007/978-3-319-69718-5_7

and industrial developments of the 1960s produced an environment open to scientific and technological relationships, and the sharing of expertise in European and international organisations; the World Health Organisation would be influential in regard to antibiotics as, eventually, would the European Communities, later the European Union. During the transition to democracy following the dictator's death in 1975, a focus on public health concerns emerged. This included the issue of drug resistance and from the 1970s onwards, national awareness campaigns were mounted, pre-empting restrictions on antibiotic consumption in European agriculture. There is an additional story to be told here, that of penicillin's agency in twentieth-century history through the trajectory of resistances, its evolving role in living bodies and public policies.[1]

'Will microbes become immune to antibiotics?' was the disturbing title of a news report published by a popular Spanish weekly in February 1958. The likelihood of some microbes becoming 'immune' or resistant to antibiotic action, and of a time in the future when antibiotics would no longer be effective against bacteria, was qualified in the report as 'the great problem of medical biology'. Resistant bacteria, according to the commentary, secreted an 'enzyme' (quotation marks in the original) that neutralised the drug's action.[2] The report issued a public warning about the chemical reaction of microbes while also emphasising the link between microbes and medicine and the pivotal role scientific research had to play in solving the problem.

Awareness about resistance to antibiotics, and the toxicity and side effects of penicillin and later antimicrobials spread rapidly. To understand how these issues became common knowledge so quickly requires examination of the extensive use of penicillin, from the anxieties preceding its marketing and administration in the late 1940s to warnings disseminated during the 1960s on overprescription and drug-resistant bacteria, in relation to the empire of consumption, increasing pharmaceutical expenses in Western nations, antibiotic use in farming and veterinary medicine, and the expansion of biological research.

Resistance did not have the same meaning in the clinic as it did in the bacteriology or microbiology research laboratory. In the clinic it meant a patient was not cured of infection, fever and fear remaining; in the laboratory it referred to the survival of bacterial culture. Some antibiotics were known to be active in vitro and not in vivo, and vice versa.[3] Among the strategies adopted by clinicians, researchers and the pharmaceutical industry to combat antibiotic resistance from the early 1950s onwards were attempts to find new antimicrobial drugs (Chap. 5). Research was also conducted into antibiotic mechanisms, and the bacteria that resisted, proving itself either partly or wholly immune to antibiotic action. A third approach was to warn against the overconsumption of these antimicrobials, attaching blame to overprescribing physicians. In the midst of a medical reform of the antibiotic era, a time when patients awaited prescription as a first step toward cure, the collection of statistical information began. The emerging epidemiology would address the incidence of infectious disease, among other phenomena, and provide statistics on antibiotic usage and the effects in a given population of either animals or human beings.[4]

In this chapter I explore these three strategies, particularly work carried out in Spain, in relation to both the political circumstances during the transition to democracy after Franco's death in 1975 and international agreements on health care and drug use. I also address antibiotics as both subjects of, and tools for, research in the biological laboratory.

Awareness of Resistance in Medical Practice

Information travelled easily. Soon after the first batches of penicillin had arrived in Spain during the 1940s, awareness of the dangers posed and overconfidence in its healing capacities prompted clinicians to warn against misuse of the drug. The biography of penicillin transited during the long period of Franco's rule from its main role as a cure against microbial infections and poverty to a commodity requiring careful administration. During that time the drug's chemical composition would be modified only slightly and combined with other, newer antimicrobials in the marketed preparations.

Penicillin, as other antibiotics would, became one of the most well-known substances in the world. Information produced during the first decade of penicillin's biography included its chemical composition and structure, described by the British X-ray crystallographers Dorothy Hopkins and Barbara Low in 1945, and the resistances it created in the clinic, analysed by British pathologist, Mary Barber. After graduating at the London School of Medicine for women, Barber studied not only the spread of *Staphylococcal* infections but performed microbiological tests to identify the strains involved. Her studies took place at both the bench and the bedside, enabling her to describe the transmission and microbiological meaning of the phenomenon in detail.[5] The existence of resistance thus became clinical knowledge in 1947, when

Barber published her research on the effectiveness of penicillin against *Staphylococci*, and the resistant strains she had identified. She discovered the spread of infection was due to a resistant strain carried in the nostrils of nurses caring for infected patients. These were, according to Barber, 'naturally resistant strains which survived by a simple process of selection in penicillin-treated infections' and such changes in 'bacterial flora take place during the course of penicillin treatment'.[6] As her colleague Lawrence P. Garrod stated after her untimely death in a traffic accident, 'she was concerned with the problem of how to make the best use of a much wider range of antibiotics in controlling hospital infections'. It was on the basis of her work, 'and energy' added Garrod, that a policy on the use of antibiotics was agreed. Restrictions were placed on its use as a prophylaxis—although it is not clear whether this strategy was adopted by all physicians or merely at health authority levels—and in an attempt to prevent the development of resistances, particularly in *Staphylococci*, it became routine to prescribe antibiotics alongside other chemicals.[7] Strikingly, it was well over a decade before the data collected by Barber in 1947 would be taken seriously: that is, that resistance was an event observed in hospital and it should be in hospital that the use of antibiotics be carefully monitored and controlled. It was not until the British parliament established the Swann Commission in 1969 that concerted efforts were made to evaluate the issue of resistance. Initially intended to focus on resistance in medical practice and the prevalence of infections in Great Britain in both human beings and animals, the Swann Report was ultimately dedicated to analysing the relationship between antibiotics in veterinary medicine and animal fodder, and the development of resistance in human beings. It was Barber's colleague Garrod who successfully campaigned for a discussion that ended by recommending the prohibition of antibiotics in food for livestock.[8]

The hospital was regarded in the UK as a safe, clean place where patients could be taken to be treated and cured.[9] In Spain during the 1950s, however, the public hospital network was as yet underdeveloped. Information about the use of antibiotics, their healing and celebrated activity against infections, had been gathered mostly by general practitioners from cases treated at home and in dispensaries, for respiratory infections and venereal diseases, with the new drugs often prescribed only after older remedies—including sulfa-drugs—had failed (see below and Chap. 4).

Salvarsan was known to be toxic, and sulfa-drugs could produce resistance. Early concerns that penicillin might also be highly toxic were unfounded. Resistance was experienced less in the initial triumphant days of penicillin in infection therapy. As a newer and thus lesser-known phenomenon, experience and training in reporting and evaluating resistance was in short supply.

The issue was taken into consideration from the very start of streptomycin usage, however, as clinicians were then becoming more and more aware of resistance. Spanish clinicians José María Segovia de Arana and Juan M. de Palacio Mateos reported in 1947 that streptomycin could rapidly produce resistant bacteria when applied alone.[10] Following treatment failure, combinations of the drug with other chemicals and antimicrobials would be administered immediately, thereby demonstrating the capacity to find solutions to the problem; at least this was the message from research laboratories and pharmaceutical manufacturing plants.[11] In 1948, streptomycin was proved effective when combined with para-aminosalicylic acid, which immediately became widely known as PAS,[12] and clinicians and medical practitioners began to publish their results.[13]

It is not clear whether Spanish clinicians read the *British Medical Journal* issue in which Mary Barber published her results; certainly the 1947 volume is missing from the collection received by the medical research institution in Madrid and, according to the general catalogue, is not in any Spanish university library—of course it may have been taken out on loan and not returned.[14] This suggests Barber's report may have gone unnoticed, although it is impossible to know if any clinicians received the journal by subscription.[15] Given the scarcity of penicillin in 1947, a year before Fleming's visit (see Chap. 2) it is likely the issue would not have appeared hugely significant to the Spanish medical profession.

In 1952, a report on the new antimicrobial terramycin, which included data on resistance, was published in Spanish.[16] Also in 1952, Carlos Jiménez Díaz—chair of the Comité Nacional—reviewed the adverse effects of a number of antibiotics. These included fungal infections, bacterial resistance and diarrhoea and intestinal infections caused by antagonism of gut flora. By that time the 'high frequency' of resistance produced by more recently marketed antibiotics, among them aureomycin, erythromycin and terramycin, was well known. According to Jiménez Diaz's review, resistant strains could be produced by mutation, as a survival of mutants; as if resistance was acquired, according to the

US-based Yugoslavian geneticist, Milislav Demerec. Jiménez Díaz cited Mary Barber's studies on *Staphylococci* resistance, and stressed both the biological and 'practical' importance of this information: 'with the routine use of antibiotics we can create resistant strains, some of them acute, or even deadly or by producing an infection of a different kind, maybe not as grave but more difficult to defeat'.[17] Bacteria were known to be everywhere, creating what would soon be termed 'ecologies'—natural environments to be preserved—in the mouth, pharynx, respiratory system and intestine, protecting the living body. A new corpus of knowledge on bacteria emerged in parallel to that of antibiotics and resistance. Bacteria were not only those microbes that created the infections sera and vaccines were used to prevent, but components of health itself. The destruction of these microbes by antibiotics created defencelessness, possibly the most serious adverse effect in the new landscape created by the extensive use of antibiotics.

The Media, Prescriptions and Public Awareness

Testimonies from medical publications and personal recollections suggest the wonder drug was mostly prescribed by Spanish doctors during home visits to their patients and, from the 1950s, for venereal disease sufferers visiting dispensaries.[18] This may account for the later detection of resistance in Spain compared to the UK and USA, where hospitals—the main settings for resistance detection, as Barber had shown—were more common. The scarcity of penicillin and delay in sufficient provision would also have been agents in the appearance and recognition of penicillin resistance. Resistance was named, although from this it does not necessarily follow that the phenomenon was known first-hand. The handling of resistant cases would have created knowledge and practices, but many resistances may not have been identified; this, of course, is difficult to verify. Medical reports from the late 1940s and early 1950s onwards suggest Spanish clinicians were aware resistance to penicillin existed, but it is not clear whether this stemmed from personal experience.

The issue of resistance began to appear in Spanish daily newspapers from 1950, one of the first times in a quote from Selman Waksman during his visit in March of that year. Arriving by train at the Spanish–French border, Waksman was greeted by a group of scientists and academic officials. After lunch in Barcelona, he continued by train to present his lectures on streptomycin in Madrid, at the main hall of the

Consejo Superior de Investigaciones Científicas and at the Academy of Pharmacy, where he was made an honorary member, and later in Seville at the Colegio de Farmacéuticos. Having been invited by the Sociedad Española de Microbiología, the lecture Waksman presented was translated into Spanish by Florencio Bustinza and published in the society's journal, *Microbiología Española*.[19] In addition to describing the discovery of streptomycin—by then renowned for its effective treatment of some types of tuberculosis—by himself and Albert Schatz, Waksman mentioned resistance, a phenomenon that had been observed when treating certain kinds of bacteria with the new drug.[20]

Despite these warnings, penicillin retained its status as a successful therapy: in 1953, a report in *Microbiología Española* by Kenneth B. Raper celebrated penicillin's tenth birthday by acknowledging the drug as one of the most effective and least toxic antibiotics available. Raper had spent the Second World War in Peoria, working on improved strains for the submerged culture of penicillium.[21]

While celebrating penicillin's success, news of resistance appeared as part of widely distributed information on the antibiotic: 'Careful, *Staphylococci* resisted penicillin', was the title of a 1955 commentary on an article by Fleming published in the UK journal, *The Practitioner*.[22] Restriction of antibiotic use to cases of 'real need'—that is, not for flu and 'sniffles'—was recommended. Then, as now, commentaries included warnings and information on how to take antibiotics correctly, and emphasised their ineffectiveness against flu. Although these warnings and reports on resistance appeared in Spanish newspapers, antibiotics continued to be sold without prescription.

In the 1940s Spanish health authorities had regulated antibiotic use by making prescription a legal requirement. Although a 1945 decree established the possibility of selling surplus amounts in pharmacies, prescriptions for these early batches would still be required, authorised by the medical association of each province, and the Comité Nacional retained control of penicillin received from abroad.[23] In January 1950, with penicillin expected from the two bottling plants, the government permitted the drug to be sold in pharmacies when available quantities allowed; this time no mention of prescription was included in the Orden (Order),[24] which indicates it was sold without prescription. Prescription was merely a means of government control over the new drug, and had nothing to do with resistance or any other health concern.

At some point the unwanted effects of new drugs in circulation were noted. An initial list of medical pharmaceuticals considered dangerous enough to warrant prescription and medical control was published in 1965. Antibiotics, however, were not among the eighteen types of pharmaceuticals listed, which included hormones, corticoids, anaesthetics and antiepileptics.[25] It was in 1977, when information included on the prescription form itself was under review, that 'anti-infectious' drugs were included, and remained subject to prescription through a 1985 review.[26]

Antibiotic usage has remained high through to the present day, and penicillins have the highest consumption figures.[27] Despite the supposed requirement for prescription, compliance has largely been absent until very recently. As stated above, however, patients did begin listening to warnings against the misuse and overconsumption of antibiotics from 1950, a time when farming was also consuming antibiotics (see below).

The global circulation of infections was regarded to be as worldwide as antibiotics. Venereal diseases were a major international concern, and penicillin already one of the most promising therapies in 1948, when it was recommended that the purest form then available be 'restricted to syphilis', and the World Health Organisation made its call for penicillin production to be expanded. In 1977 the WHO published its first report on 220 'essential drugs'. The category 'anti-infective drugs' contained a number of sulfa-drugs and 16 antibiotics, including ampicillin, penicillin and some of its derivatives, and chloramphenicol.[28] Intended to address 'the main drug problem facing developing countries'—a lack of pharmaceuticals—the 1977 report also outlined 'possible new drug policies'. Expenditures on drugs accounted for around 40% of the total healthcare budget in developing countries, but governments of 'affluent nations' were also becoming concerned about drug costs. In tune with the concerns of clinicians and biomedical researchers on the role of overdosing and overprescription—regarded as the main agents in the development of resistance to antibiotics—the report recommended 'rational prescribing'. For European countries, most of which had a national health service providing medical care and covering the cost of drugs prescribed by public service doctors, the rising cost of medicines was of great concern; this included Spain, where the state covered—as it still does—part of the cost of drugs prescribed through the national health system.[29]

During the late 1970s and early 1980s, penicillins were consumed in vast quantities on a global scale. Amino-penicillin was a particularly widely consumed antibiotic in Spain between 1976 and 1983. Although

I could not find figures on earlier consumption, CEPA annual reports mention increasing sales of penicillins and streptomycin from the early 1950s and throughout the 1960s. As Anne Kveim Lie has stated, during the 1970s 'the old penicillins were still popular with the [Norwegian] health authorities', although doses and schedules had been significantly altered. Norwegian health officials were sceptical of the new, once a day dose,[30] and continued to rely on the older, cheaper penicillins which, as would later be discovered, were less likely to provoke resistance. This Norwegian policy was a pioneering warning about antibiotic resistance, and has had international influence on attempts to curb the culture of overprescription since the 1960s.[31]

Research in Microbiology and Resistance

In 1953, Spanish microbiologist Antonio Portolés published his first study on resistance.[32] As a microbiological problem generated by the success of penicillin and other antibiotics, the process of bacteria adapting to a molecule attacking its growth, or destroying its membrane, was a fascinating biological event. Portolés worked as both an army pharmacist and assistant in the laboratory of microbiology at the Centro de Investigaciones Biológicas, Madrid, where he gained his PhD and continued his research until retirement.[33] When combined therapies were being developed to avoid resistance, he carried out in vivo research on these combinations. The history of resistance, or a part of it, includes the collection of resistances: collection of the morphologies bacteria adopt when antibiotics are added to a culturing media. The natural history of microbes belongs to the long history of mycology, strain collections and classifications: names and databases that significantly predate the era of big data. Amateurism and disciplinary, academic expertise in collecting fungi and preserving samples composed the style of knowledge and practices to which antibiotic research belongs.

Awareness of the multiplicity of morphology that bacteria could adopt in the presence of antibiotic substances led Portolés to advocate careful culturing of microbes involved in clinical infections and in vivo testing of their action. Portolés and his colleague, the biologist Teresa Pérez Ureña, developed an enduring and productive research group in Madrid, adopting new techniques as they became available for the study of resistance and antibiotic action, observed in Petri dishes and later in ribosomes.[34] During the 1950s and 1960s, resistance began to be associated with bacterial mutations, the mutant products of modified bacteria which exhibited antibiotic resistance; therefore, during the nucleic acid era, ribosomal RNA became the cellular unit for the study of antibiotic action in molecular microbiology. Among the most productive researchers in Spain from the 1960s onwards was David Vázquez.[35] On his return from England in the mid-1960s, Vázquez carried out research on the action of antibiotics in protein synthesis inhibition, maintaining the research agenda he developed with Ernst Gale at the University of Cambridge. Microbes had become experimental systems: *Escherichia coli* and many others were adopted by a set of laboratories during the 1960s. Their nucleic acids and protein biosynthesis were studied during a golden age of molecular biology and research in general. The pulse and chase experiment—in which radioisotopes were used to detect and follow RNA biosynthesis, the incorporation of a molecule into a protein or nucleic acid—became an extremely useful method. Penicillin and later antibiotics joined in as productive tools for studying bacterial growth and ribosomal RNA, as utilised by David Vázquez, demonstrating their roles as agents in the history of molecular and cell biology. Not only was research required to learn about antibiotics' modes of action, their activity at the level of the cell and its molecules played a significant part in the production of biomedical knowledge during the 1960s and 1970s.

Antibiotic Consumption and Health Reform

In her 1985 study, the infectious disease expert Beatriz Pérez Gorricho carried out research into antibiotic usage in Spain over the previous decade. By the end of 1975, at the start of Gorricho's time period, Spain was no longer a poor country governed by a dreaded dictator: Franco had died in his bed, still *Jefe de Estado* (Chief of State), on November 20. While many Spanish citizens came to Madrid to pay their respects, a long list of foreign prime ministers and heads of state came to celebrate the coronation of King Juan Carlos instead, held in one of the oldest and largest churches in Madrid, Los Jerónimos; despite the power of the Catholic Church, Madrid lacked a cathedral throughout the dictatorship. From that time forward, social and health policies were part of negotiations for a peaceful transition to democracy, included in the manufacturing of a political and economic agreement beyond ideologies. The quantification of antibiotic resistance, as part of an evaluation of antibiotic use in general, slowly became integrated into public health reform and prescriptions became compulsory.

The political agreement known as Pactos de la Moncloa, signed by representatives of all political parties in the first democratic parliament, was initially deemed necessary to overcome the economic crisis it was feared might hinder the democratic transition and establishment of democratic institutions. Among the ten points dealt with was reform of the social security system, which included medical care, preventive medicine and the regulation of some drug use. Without mentioning any particular pharmaceuticals, it suggested a revision of their use and consumption.[36]

The Pactos de la Moncloa made no mention of research.[37] The development of biomedical research in the West—which had benefited from increased funding for scientific and technological research by the USA and UK since Sputnik I had been launched by the Soviets in 1957—had been a strong influence and powerful reference for the modest but highly productive and ambitious scientific research community in Spain. Bacterial genetics and microbiology developed in an international network of knowledge and practices that combined a natural history approach—the collection and classification of specimens—with biochemical and molecular experimentation.

Since the 1960s, the expanding Spanish community of biomedicine—including microbiologists, cell biologists and bacterial geneticists—was composed of biological researchers and clinicians who combined medical practice and research at the bedside with the clinical laboratory. From the 1960s onwards, sharing in Western economic and industrial development, many would become highly productive researchers and science policymakers. This large group of researchers from university laboratories and the research institution CSIC, became one of the bases upon which an influential scientific community grew in expertise and impact.[38] After the oil crisis during the late 1960s and early 1970s that affected the whole of Spanish society, biomedical research benefited from the more open international relations the democratic regime secured. Spain joined the European Communities in 1986 and was therefore able to benefit from European science policy, particularly by joining institutions in Brussels and participating in the projects funded by successive Framework Programs.[39]

According to Pérez Gorricho, antibiotic use in Spain decreased slightly between 1975 and 1985 from 31 daily doses per 1000 inhabitants to 28, while total expenses decreased from almost 14 million constant pesetas to 10.5 million. During that period, penicillin, including aminopenicillin—the first semi-synthetic penicillin developed in the UK in the 1950s—was the most commonly used.[40] In 1982, Spain consumed more antibiotic units than any neighbouring country, a total of 82,754 units per thousand inhabitants, followed by Italy at 75,606 and France at 72,364 units. Penicillin was by far the most frequently used: in 1983 more than 1000 prescriptions per thousand inhabitants were for broad-spectrum, reduced-spectrum or 'general' penicillins, while the figure for cephalosporin, tetracycline and others was 350 prescriptions.

One of the effects of antibiotics is growth promotion, hence their popularity as additives in animal fodder. The possibility of preventing infection in the increasingly industrialised and overcrowded agriculture of Western societies since the 1950s also seemed beneficial. By the time of Pérez Gorricho's study, Germany had prohibited the use of antibiotics in fodder,[41] and the British Swann Report had recommended penicillins and tetracyclines be prohibited as fodder additives in the UK during the late 1960s.[42]

Pérez Gorricho suggests a structure of antibiotic consumption in Spain between the 1970s and 1980s divided into medical—veterinary and human—and non-medical use in fodder. At that time antibiotics not marketed for 'application in human pathology' were permitted as fodder additives. In Spain in 1984, according to her estimations, veterinary medicine consumed between 30 and 35 tons, human medicine between 350 and 400, and farming a minimum of between 226 and 400 tons. Spanish livestock was consuming a similar number of doses of antibiotics as civilians.[43]

Restrictions on fodder components were introduced in Spain in 1988. Although antibiotics were not banned, manufacturing conditions were regulated and detailed labelling of composition enforced.[44] In previous regulations 'antibiotics' had not even been mentioned.[45] Since 1996, additional regulations have further restricted the use of antibiotics in fodder. It was in 1999, following research-backed appeals from Scandinavian countries, that the European Union banned the use in fodder of the antibiotics zinc-bacitracin, Spiramycin, Virginiamycin, and Tylosin phosphate.[46]

In regard to the settings where penicillin and later antibiotics were prescribed and consumed in Spain, only six per cent of antibiotics were consumed in hospitals during the 1980s. Total expenditure had decreased 25% since the mid-1970s. An increase in the number of medical practitioners and clinicians of 41% at the same time as the prescribing of antibiotics fell significantly, suggests improved policies in regard to healthcare services, pharmaceutical prescriptions and expenses taking place from the early 1980s onwards.[47] The infections most likely to warrant the prescription of antibiotics were those of the respiratory tract. According to Gorricho, the relatively high amount of antibiotics consumed in Spain at that time was due to a higher prevalence of infections than neighbouring countries. But, she added, antibiotics were still incorrectly prescribed in 50% of hospital cases, either as a prophylaxis or to prolong treatment, thereby eliciting resistance and increasing the prevalence of infectious disease.[48]

These were the years of national healthcare reform in Spain, with new primary care centres created across the country, and the right to healthcare extended to the entire population. Although the number of antibiotic prescriptions decreased during this period, total expenses on antibiotics remained constant, according to Pérez Gorricho. The ecology of resistance, however, included a less publicly known agent: the concentration of antibiotics in residual water on farms and in human environments was increasing. As greater amounts of antibiotics participated in the microbial flora of animals and people, infection originated from the apparently clean environment of the rich, well-developed society Spain was then becoming. All these conditions together were regarded by Perez Gorricho as participating in the structure of resistances.

Medical practitioners played a part as agents in this ecology, as did the promise of a disease-free world, despite the continued emergence of new infectious diseases. The economic development of the 1960s provided the socioeconomic background that suffered during the oil crisis. But the political promises of democracy also had a part to play by making discussions on public health as a public issue possible, and by collecting sources and data about the prevalence of infections to be passed onto health authorities: the emergence of epidemiology.

The creation of new large hospitals was a product of the development and consolidation of the 1960s. During the late 1970s there was an increase in the number of practitioners and intensifying enthusiasm for public health policies, in Spain and internationally.[49] The much-criticised model centred on responding to illness would begin to shift to the promotion of health and disease prevention, championed in widely distributed internationally influential reports including the Lalonde Report in 1974, the Alma-Ata Conference on primary care in 1974 at which ‘Heath for all in 2000’ was proclaimed, and the Ottawa Letter of 1986.[50] The extension of medical care to all Spanish citizens would be regarded a part of this environment, which combined the expectations created by democracy with international public health promotion.

The first post-Franco government created a Ministry of Health in 1978; healthcare had previously shared a department with policing, suggesting both crime and illness were regarded by the Franco regime as matters of ‘social order’. The Socialist Party, voted to power in 1982, began a political strategy known as *Reforma Sanitaria* (Sanitary Reform) resulting in the 1986 *Ley General de Sanidad* (The General Health Law).[51] It was during these first decades of democracy that the issue of antibiotic resistance began to be addressed at policy level, in part a consequence of the higher political importance health and medical care were afforded, now with a dedicated minister attending weekly cabinet meetings with the prime minister. A broad consensus for healthcare reform was being negotiated between medical practitioners and the government. Although budgets would always remain smaller than expected, healthcare as a national political system was prioritised.

Infections would never disappear, as old microbial foes re-emerged in an overcrowded urban society continually stretching resources. Penicillin and the other industrially manufactured products of the biomedical research system of the 1950s helped create the era of consumption.[52] Although the promise of penicillin would never be fully realised, it participated in the long post-Second World War era, through to the turn of the century, precisely by creating the expectation that prompted the rise of resistance.

Since the 1950s, the Spanish community of microbiologists had become increasingly involved in the new cellular biology research agenda. Focused on bacteria and its morphology, explanations were sought in biochemical terms, and the issue of resistance was carefully analysed in bacterial cultures by researchers such as the groups led by Teresa Pérez-Ureña and Portolés.[53] Although there was disagreement on the extent to which antibiotics would become harmful, a general agreement regarding the danger they posed did prevail. Antibiotics tested in milk were found responsible for creating resistance and delaying the production of cheese. The activity of chloramphenicol in bacteria was shown to affect the biosynthesis of protein. The association of antibiotics and fungal infections in patients was studied through experiments performed in vitro, in the Petri dish.[54] An entire research agenda developed on the characterisation and classification of bacteria and fungi by their reaction to lytic substances: those able to break the bacterial wall or 'envelope'.[55]

Research was not solely focused on resistance. Antibiotics played an influential role in the study of microbes' molecular biology: how they went through the bacterial wall, entered the cytoplasm and produced reactions, intervening in nucleic acid biosynthesis. The Japanese microbiologists, Tsutomu Watanabe and his colleagues, reported that resistance development was due to what they termed R factors, opening a new research line that would also enrich research in bacterial genetics. Named plasmids by Esther and Joshua Lederberg, these hereditary units were described by Watanabe as having the property of carrying resistance and the ability to disseminate among bacterial populations. Watanabe reported the detection of R Factors when investigating cases of multiple resistances in *Shigella* strains, one of the leading causes of diarrhoea in Japan and worldwide.[56] Resistance proved to be multifactorial. Plasmids, as cytoplasmatic genetic units, became the main threat, thus geneticising the process while adopting the terminology of molecular biology, and placing nucleic acids, of both the cytoplasm and nucleus of the cell, as the focus for explanations of biological activity and resistance.

Antibiotic Consumption, Resistance and the Laboratory

The landscape of antibiotic practices had expanded to include the consulting room, the hospitals, the laboratory and, most importantly, the bodies that developed resistance and for which no immediate cure could be found. Since the early work by Emmy Klieneberger-Nobel, Milton Salton and by Esther and Joshua Lederberg, much research had been carried out on the changing morphology of microbes in the presence of penicillin and other antibiotics. Microbiologists exploited resistances, using antibiotics not only as research subjects but as tools to study bacteria and microbes in general. Throughout the decades of antibiotic isolation and therapy, great quantities of new drugs were described, tested, industrially manufactured and marketed by the pharmaceutical industry. This antibiotic abundance participated in the irresistible empire of consumption promoted from the USA, the expansion of which contributed to the new medical culture that characterised civilisation during the second half of the twentieth century.

Resistance to the action of antibiotics was noted and described very early on. This was an old phenomenon, active against sulfa-drugs since the early 1930s, and the fact that antibiotics could produce changes in a microorganism and no longer be active against the infections it produced was known. It had also been known that some strains were resistant since the early use of penicillin, and that these survivors of penicillin action would reproduce in a new environment as non-resistant strains decreased. Microbiology and the clinic interacted at least partially, exchanging information and warnings, while industrial production appears to have acquired a life of its own, particularly once the effects of penicillin on animal growth had been noted and an entire new market opened up. Antibiotic production and healing activity marked, and indeed labelled, an entire period of the 'therapeutic revolution'—the antibiotic era—and interacted with the genetisation produced by an increasing focus on genetic disorders and congenital traits, also sought for in bacteria. The emergence of bacterial genetics would rejuvenate and develop the old debate about adaptation and mutation as competitive processes in the resistance phenomenon.[57] The changes produced in a microorganism would then be interpreted in genetic terms, with demonstrations of the genetic mutations resistant microorganisms had undergone.[58]

The activity of antimicrobials, as witnessed with penicillin, streptomycin and the many that followed, including chloramphenicol and erythromycin, proved a double-edged sword. Overuse was argued to be at the origin of resistance, while inactivity was an immediate event, observed the first time a drug was administered for a microbe it was inactive against. Trust in scientific progress produced a future full of expectations that drugs active against every troublesome microbe would be found and manufactured. Research into the microbiology, genetics and biochemistry of antibiotic action and resistance would develop far earlier than the heeding of clinical regulations and warnings by clinicians, health authorities and the public.

In the early 1980s, it was estimated that 41% of hospital inpatients were prescribed antibiotics, most often penicillins: levels of penicillin consumption per capita in the USA and Spain were very similar. It was in outpatient consumption that Spanish levels were significantly higher than neighbouring countries. According to Pérez Gorricho, this was in large part due to the dramatic expansion of the Spanish health service and the inadequate provision of primary care that would have controlled private consumption more efficiently.

While medical overuse was considered a major factor in the development of resistance, antibiotic use in livestock farming took also part in the bacterial ecology of resistant transmission.[59] After the Second World War meat would become increasingly desirable and affordable. In promoting growth, penicillin and many other antibiotics were participating in 'the consumer society', creating bigger and apparently safer animals for human sustenance. Industrial 'factory' farming, with animals reared in small spaces, protected from the increased risk of infections by preventive antibiotic treatment, also participated in producing resistances in animal intestines. Socio-economic changes took place and played their part in the reduction of infections, in which drug trajectories would seem to have played a significant part. The powers of health and the wealth that pharmaceutical firms provided also participated in the growing authority of the laboratory as the place where new cures would be found and resistances detected. Medical practice appears to have played a less influential role: prescription was not compulsory for the purchase of antibiotics and their addition to animal fodder became one of the main challenges in public health regulation.

Notes

1. This chapter does not pretend to be an exhaustive reconstruction of the history of antibiotic resistance in Spain, but rather a reflection on the cultures of antibiotics and on the public and professional awareness the phenomena of antibiotic resistance produced. The history of resistance in Spain deserves to be fully researched. A well-recognised reconstruction of the situation in the USA appears in Scott Podolsky, *The Antibiotic Era: Reform, Resistance, and the Pursuit of a Rational Therapeutics* (Baltimore: Johns Hopkins University Press, 2015).
2. Emilio Novoa, 'Los microbios llegarán a hacerse inmunes a los antibióticos', *Blanco y Negro* 22 (1958): 109.
3. Anne Kveim Lie, 'Producing standards, producing the Nordic region: Antibiotic susceptibility testing, from 1950–1970', *Science in Context* 27(2) (2014): 215–48.
4. Jeremy Green and Elisabeth Watkins, *Prescribed: Writing, Filling, Using, and Abusing the Prescription in Modern America* (Baltimore: Johns Hopkins University Press, 2012).
5. L. P. Garrod, 'Mary Barber. 3 April 1911–11 September 1965', *The Journal of Pathology and Bacteriology* 92(2) (1966): 603–10.

6. Mary Barber, 'Staphylococcal infection due to penicillin-resistant strains', *British Medical Journal* 2(4534) (1947): 863; Mary Barber, 'Coagulase-positive staphylococci resistant to penicillin', *The Journal of Pathology* 59(3) (1947): 373–84.
7. Garrod, 'Mary Barber', 605.
8. Bud, *Triumph and Tragedy*, 181.
9. Condrau and Kirk, 'Negotiating hospital infections'.
10. See the synthetic historical reconstruction by William C. Summers, 'Microbial drug resistance: A historical perspective', in *Bacterial Resistance to Antimicrobials*, eds. Richard G. Wax, Kim Lewis, Abigail A. Salyers and Harry Taber (Boca Raton: CRC Press, 2008), 1–9. Regarding Spain, see José M. Segovia de Arana and Juan M. de Palacios Mateos, *Estreptomicina* (Madrid: Paz Montalvo, 1947), 67–74.
11. A. Bravo Ortega, R. Pueyo García and A. Martínez Martínez, 'Nuestra experiencia en el tratamiento de la meningitis tuberculosa', *Revista Clínica Española* 51 (1953): 397–411; L. Cifuentes Delatte, 'Tratamiento actual, médico y quirúrgico, de la tuberculosis del aparato urinario', *Revista Clínica Española* 53 (1954): 281–303.
12. Christoph Gradmann, 'Re-inventing infectious disease: Antibiotic resistance and drug development at the Bayer company 1945–1980', *Medical History* 60 (2016): 155–80, on 162. See also H. W. Florey et al., *Antibiotics: A survey of Penicillin, Streptomycin, and other Antimicrobial Substances from Fungi, Actinomycetes*, Bacteria, and Plants (Oxford: Oxford University Press, 1949).
13. L. Pau-Roca and J. Coll-Sampol, 'Terapéutica antibióticas, bacteriológicamente dirigida, en asmología', *Revista Clínica Española* 66 (1957): 313–22. Carlos Jiménez Días, 'Significación conceptual del conocimiento de la endocarditis bacteriana', *Revista Clínica Española* 70 (1958): 277–9.
14. I have checked at the catalogue REBIUN, which encompasses the catalogues of all Spanish universities and research institutions, available at www.rebiun.org/catalogoColectivo/Paginas/default.aspxm, accessed April 6, 2017.
15. Although most of those who subscribed to journals would donate their collections, some libraries no longer accept hard copies if available online.
16. A. Ortega Núñez, 'Terramicina (un nuevo antibiótico)', *Revista Clínica Española* 40 (1951): 332–9.
17. Carlos Jiménez Díaz, 'Observaciones sobre el empleo de los antibióticos', *Revista Clínica Española* 47(3) (1952): 143–53, on 147.
18. Ramón Castejón Bolea, 'Access to penicillin for the treatment of syphilis in Spain in the 1950s: The role of the international agencies', *European Social Science History Conference, Valencia 30th March 2016*; Ramón Castejón Bolea and Esteban Rodríguez Ocaña, 'El control de las

enfermedades venéreas en España durante la década de 1950: los inicios del Proyecto España 8 de la OMS', *Congreso de la Sociedad Española de Historia de la Medicina*, San Feliu de Guixols, Girona, June 2017.

19. Selman Waksman, 'Investigación de nuevos agentes quimioterápicos', *Microbiología Española* 3 (1950): 83–97.
20. *ABC*, March 16, 1950, 29.
21. R. H. Burris and E. H. Newcomb, 'Kenneth Bryan Raper', *Biographical Memoirs* 60 (1991): 251–72.
22. 'Ojo, que los estafilococos resisten a la penicilina...', *ABC Sevilla*, January 7, 1955.
23. *Boletín Oficial del Estado* 4 de enero 1947: 'Orden por la que se autoriza la venta libre de aquellas cantidades de "penicilina" que no se considere indispensable para las atenciones de este Servicio, tanto en Madrid como en los depósitos instalados en provincias.'
24. *Boletín Oficial del Estado* 28 enero 1950: 'Orden de 24 de enero por la que se autoriza en el Territorio Nacional del medicamento Penicilina en todas las farmacias'.
25. *Boletín Oficial del Estado* 28 septiembre 1965: 'Orden de 14 de agosto de 1965 por la que se dictan normas para la dispensación de medicamentos'.
26. *Boletín Oficial del Estado* 31 de mayo 1977: Ministerio de la Gobernación, 'Orden de 11 de mayo de 1977 sobre receta médica'. *Boletín Oficial del Estado* 16 de noviembre de 1985: Ministerio de Sanidad y Consumo, 'Orden por la que se determinan los medicamentos de utilización en medicina humana que han de dispensarse con o sin receta.'
27. Ministerio de Sanidad, Politica Social e Igualdad: *Uso de antibióticos en España*. Madrid: Agencia Española de Medicamentos y Productos Sanitarios, 2010.
28. World Health Organisation, *The Selection of Essential Drugs* (Geneva: WHO, 1977).
29. See Félix Lobo, *Medicamentos: Política y Economía* (Barcelona: Masson S. G., 1992). Although not focusing on antibiotics, as an economist and policymaker during the years of healthcare reform, Lobo provides a comprehensive overview of policies on medicine use.
30. Lie, 'Producing standards'.
31. In addition to Lie, see Christoph Gradmann, 'Sensitive matters: The World Health Organisation and antibiotic resistance testing, 1945–1975', *Social History of Medicine* 26(3) (2013): 555–74.
32. Antonio Portolés, '¿Intervienen los antibióticos en la formación de levaduras?' *Medicamenta* X (1953): 53–5.
33. For biographical data on Portolés see 'Sesión necrológica en homenaje al Excmo. Señor Don Antonio Portolés Alonso', *Anales de la Real Academia Nacional de Farmacia* 72 (2006): 165–92.

34. 'Sesión Necrológica'. I could not find any biographical information on Teresa Pérez Ureña: she was Portolés's wife and long-term colleague.
35. On David Vázquez, see M. J. Santesmases, 'National politics and international trends: EMBO and the making of molecular biology in Spain', *Studies in History and Philosophy of Biological and Biomedical Sciences* 33 (2002): 517–31; M. J. Santesmases, *Entre Cajal y Ochoa* (Madrid: CSIC, 2001), 150–2. His biographical data are detailed in Miguel Somovilla, *Retrato de David Vázquez 1930–1986. De la sidra a los antibióticos* (La Felguera: Comisión Homenaje a David Vázquez, 1989).
36. *Los Pactos de la Moncloa.* Texto completo del acuerdo económico y del acuerdo político. *Madrid, 8–27 de octubre de 1977* (Madrid: Presidencia del Gobierno, 1977). Available at www.mpr.gob.es/servicios2/publicaciones/vol17/, accessed April 2, 2017.
37. Luis Sanz-Menendez, *Estado, ciencia y tecnología en España: 1939–1997* (Madrid: Alianza, 1997).
38. Jordi Camí Morell, M. A. Zulueta, M. Teresa Fernández, María Bordons and Isabel Gómez Caridad, 'Producción científica española en biomedicina y ciencias de la salud durante el periodo 1990–1993 (Science Citation Index y Social Science Citation Index) y comparación con el periodo 1986–1989', *Medicina Clínica* 109 (1997): 481–96.
39. Emilio Muñoz y Florencio Ornia, *Ciencia y tecnología: una oportunidad para España* (Madrid: Ministerio de Educación y Ciencia-Aguilar, 1986).
40. María Beatriz Pérez Gorricho, 'Farmacovigilacia de los antibióticos. Estructura de consumo y mecanismos de control' (PhD diss., Madrid: Universidad Complutense, 1985), pp. 35–57. A brief introduction to the issue of resistances is reported by Robert Bud in E. M. Tansey and L. A. Reynolds, *Post-penicillin antibiotics: From Acceptance to Resistance*, Witness Seminar (London: Wellcome, 2000), 4–5.
41. Ulrike Thoms, 'Between promise and threat. Antibiotics in foods in West Germany 1950–1980', *NTM Zeitschrift für Geschichte der Wissenschaften, Technik und Medizin* 20(3) (2012): 181–214, on 192.
42. Bud, *Triumph and Tragedy*, 182.
43. Pérez Gorricho, "Farmacovigilancia", pp. 50, 69, 76.
44. *Boletín Oficial del Estado* 28 de marzo de 1987, Real Decreto 418/1987 de 20 de febrero, 'Sobre las sustancias y productos que intervienen en la alimentación de animales'.
45. *Boletín Oficial del Estado* 23 de abril de 1975: Decreto 850/1975 'Por el que se establece la reglamentación de las sustancias y productos que intervienen en la alimentación de animales'.
46. Lie, 'Producing standards'.
47. Pérez Gorricho, 'Farmacovigilacia', p. 233. *Boletín Oficial del Estado* 29 de abril de 1986: Jefatura del Estado, Ley 14/1986, General de Sanidad.

48. Pérez Gorricho, 'Farmacovigilacia', 235.
49. Ferrán Martínez Navarro and Esteban Rodríguez-Ocaña, *Salud pública en España: de la Edad Media al siglo XXI* (Granada: Escuela Andaluza de Salud Pública, 2008).
50. On primary care as an international agreement see *Health for all beyond 2000. Declaration of Alma-Ata. International conference on primary health care*, Alma-Ata, USSR, 6–12 September 1978. At www.who.int/hpr/NPH/docs/declaration_almaata.pdf. For a discussion on the declaration by a Spanish sociologist of medicine, see Vicente Navarro, 'A critique of the ideological and political positions of the Willy Brandt report and the WHO Alma Ata Declaration', *Social Science and Medicine* 18 (1984): 467–74.
51. *Boletín Oficial del Estado* 29 de abril de 1986, Ley General de Sanidad.
52. See epidemiological reports on the prevalence of infections and other illnesses provided by the Ministry of Health annually.
53. María Teresa Pérez-Ureña, 'Antibioticorresistencia in vitro de estafilococos tratados por asociaciones de penicilina con neomicina, novobiocina y kanamicina', *Microbiología Eespañola* 19 (1966): 287–98; María Teresa Pérez-Ureña, 'Valoración de la antibioticorresistencia de bacilos gramnegativos sometidos a interacciones antibacterianas de carácter sinérgico', *Microbiología Eespañola* 19 (1966): 299–320.
54. Antonio Portoles and Enrique Feduchi, 'Micosis urinarias subordinadas a la presencia de sustancias antibióticas', *Microbiología Española* 12(3) (1959): 199–242; Genoveva Tejerina, 'Desarrollo de la resistencia a penicilina, estreptomicina y aureomicina en Mycobacterium phlei y Escherichia coli', *Microbiología Española* 13 (1960): 393–408.
55. Emilio Muñoz Ruiz and Julio Rodriguez-Villanueva, 'Un método rápido y sencillo para el aislamiento de nuevos microorganismos productores de antibióticos', *Microbiología Española* 13 (1960): 409–14; E. Muñoz, M. Victoria Elorza and J. Rodríguez-Villanueva, 'Clasificación de un microorganismo de acción lítica como variedad de Streptomyces violaceus', *Microbiología Española* 17(4) (1964): 195–208; R. López, 'Comportamiento de algunas especies de azotobacteriáceas frente a los antibióticos', *Microbiología Española* 20 (1967): 167–90. R. López, E. Ronda and A. Portolés, 'Variaciones en la composición de aminoácidos en especies de azotobacterias antibióticorresistentes', *Microbiología Española* 20 (1967): 203–18.
56. Angela N. Creager, 'Adaptation or selection? Old issues and new stakes in the postwar debates over bacterial drug resistance', *Studies in History and Philosophy of Biological and Biomedical Sciences* 38 (2007): 159–90. Tsutomu Watanabe, 'Infective heredity of multiple drug resistance in bacteria', *Bacteriological Reviews* 27 (1963): 87.

57. Creager, 'Adaptation or selection'.
58. Podolsky, *The Antibiotic Era*; Gradmann, 'Sensitive matters'.
59. Pérez Gorricho, 'Farmacovigilacia de los antibióticos'; Baquero, Martínez and Carvajal used 'ecology' to conceptualise the landscape of resistance and account for plasmid participation. See Fernando Baquero, Manuel Martínez Ferrer and Ana Carvajal, 'Ecología de la transmisión de resistencias plasmídicas a los antobióticos en el intestino humano', in *Aspectos actuales en la transmisión de la información genética de microorganismos*, eds. Antonio Portolés and Fernando Baquero (Madrid: Sociedad Española de Microbiología, 1975), 267–99.

CHAPTER 8

Penicillin in Spain, 1940s–1980s: Circulating Health, Research and Gender

My aim with this book is to present episodes in the circulation of penicillin as a scientific, medical and industrial material, in the context of Spain, the Spanish government and public, industrial manufacturing, the research laboratory and the clinic. This is a story of the antibiotic age; of a new commodity investigated through an elaboration of the meaning of its journeys and movement, utilising circulation as an analytical category. Articulating my reconstruction around a scientific object enables the introduction of a set of agents involved in the movement of penicillin, or the set of penicillins in all salt forms, between a diversity of settings. These agents participated in this historical reconstruction as they thought of, talked about, wrote, touched and experienced the action of the new drug. This is an attempt to tell a history of Spain through a scientific and medical object, as a historical agent, its shifts and transits, and its agency in culture and society beyond industry and medicine. It is the history of a culture of consumption, of a therapeutic commodity and of healing.

In this chapter I bring together all the agents previously considered separately, and attempt to provide a global approach to the circulation of penicillin in Spain. I present a conceptual elaboration of circulation while narrating a comprehensive synthesis of the early history of penicillin in Spain, reflecting on the journeys of objects across times and geographies, between the laboratory, the factory and the clinic, in the recent history of biomedical knowledge and practices. Besides and beyond the customary main characters, generally male—the pioneers, founders, inventors,

M. J. Santesmases, *The Circulation of Penicillin in Spain*, Medicine and Biomedical Sciences in Modern History,
https://doi.org/10.1007/978-3-319-69718-5_8

even the owners of rights and factories—in this story I retrieve a wide range of agents whose actions contributed to the role played by penicillin in contemporary history. By focusing on gender, women as agents—workers, researchers, technical staff, smugglers and participants in illegal trade—are retrieved, not only in Spain but also in penicillin's early days in Oxford and many pharmaceutical factories. Research practices, industrial tools, devices and experimental methods also proved to be agents.

The story begins during the first decade of Franco's repressive dictatorship, a time when many international organisations were being established. Global and national encounters and exchanges account for events in Spain, during which penicillin became a public good, a needed commodity, and a basic medical drug, even if sometimes with limited therapeutic activity. Doctors, researchers, smugglers, women in the factory line and the research laboratory, and the public participated in manufacturing a culture of healthcare that included the cure of infections, and faith in the power of science and medicine: research would provide new and better drugs, it was believed, to combat practically any illness.

The drug in my reconstruction was particularly celebrated, a new antimicrobial, the action of which produced astonishment and wonder among the public, physicians, scientists and industrialists. It was an active object, an agent that attained its own life. Produced at the bench in a Petri dish, reports of this miraculous new substance were published in a small number of scientific articles and penicillin began to travel: this scientific object circulated widely, constructing diverse social lives in the various settings it reached.

As a living machine, the mould provided the drug: a microorganism reacting to the medium in which it would be cultured and nurtured.[1] This cultivation produced a crop, and the agricultural metaphor is a useful one.[2] Living machines would be fed, observed and selected, some of the biosynthesised product isolated, purified, identified and described as a molecule: that of penicillin. The earliest travels of penicillin were of two kinds, between laboratories, and from laboratories to the testing bedside. From those early tests, the product moved to US factories.

This book has focused on later events, after the Second World War, when knowledge about penicillin had circulated and made the drug a popular desire: the chosen material to fight infection. The earliest batches in Spain were sent from the USA, and replaced by bulk amounts following the signing of agreements with Merck and Schenley for domestic manufacture. I have focused on Merck and its Spanish partner,

CEPA. The methods for manufacturing penicillin travelled from Merck headquarters at Rahway to Madrid, where CEPA constructed a bottling plant and factory for penicillin production under license. Instructions on how to build the factory and handle devices for producing penicillin were included in the agreement, kept secret until patented in Spain.[3] The drug was produced at the end of a bottling plant factory line, where women participated in the manufacture of a purified, purifying substance. Such a gendered industrial space brings this narrative to women's place in the research laboratory and the industrial world, and to the gendered object penicillin had become in the assembly line.

Legal and black markets coexisted: as the drug remained scarce and expensive, costing far more in Spain than the USA, where it had become relatively cheap by the end of the Second World War, this coexistence lasted over a decade.[4] Smuggling and illegal trade played a part in the construction of penicillin's public renown, curing those patients able to pay for black market penicillin, at a price subject to the law of supply and demand, the root of capitalism. Secret journeys were undertaken, in public and private transport, while penicillin was prescribed, bought and consumed, legally or not.

Penicillin, and later other antibiotics, travelled fast, through an underexplored trajectory in the historiography of knowledge and manufacturing. This speed was comparable to that of infections. Almost as swiftly resistance arrived, as a biomedical concept and a physical experience at the origins of epidemiology. As the phenomenon of resistance to penicillin also circulated, microorganisms were increasingly surveyed for new drugs to face those infections penicillin would not cure. It was research carried out on the action of penicillin, both in Spain and abroad, which provided new insights and new substances. One of these resulted from the screening programme established by CEPA and Merck in Madrid; a successful antibiotic produced by this programme, fosfomycin, is still on the market. With the research programme travelled the promise of post-war recovery based on research and industrial production, fully intertwined in the policies of international agencies created after the Second World War. Following a decade of training a new screening method enabled the detection of other antibiotics and medicines. Overconsumption by clinical patients and farm animals provided more insights. Decades of cure coexisted with decades of overconsumption and resistance.

Circulation: Geography and Time

Besides its popular meaning, circulation is a term with postcolonial claims; it problematises travels by reflecting on the details of such journeys, to pose and eventually answer questions regarding how knowledge and practices travel through time and geography. These movements are not taken for granted here, but included in the account of penicillin as an agent: circulation is itself history.

This narrative is a history of the circulation of both knowledge and a particular commodity. It is about how, in addition to cure, circulation created symbols of power, political, scientific and therapeutic, promises that the devastation of civil war and the miseries of everyday life under Franco during the 1940s would be overcome. Inspired by Maria Rentetzi, I would say that circulation includes here a reconstruction of a trafficked material, a commodity in motion.[5] As has been discussed, knowledge about penicillin circulated earlier than the drug itself, spreading through society by what Robert Darnton has termed communication circuits. An object runs a full cycle through a communication circuit, each part making sense when related to the whole: when embedded in a wider history.[6] From its early isolation to reaching the infected wound, after interacting over space and time with the set of systems that composed the world in which penicillin circulated, the product and its antibiotic action returned to the laboratory, where its behaviour in the cell was explored. Penicillin returned to the professional domain where the new antimicrobial had been detected and isolated for clinical testing. The drug recovered its place as a scientific object, having fully obtained its medical agency, as a therapeutic, marketed commodity.

This story is a lengthy one, involving a diversity of distant agents. Distance, a physical separation in both time and space, was overcome by the fact the places of departure and arrival, though far apart, shared a social culture of medicine, trade and science, and also of travels as sources of knowledge. Penicillin circulated within the West—although Japan, with its long history of fermentation and established links with the West should be included in that metropolitan area—and between political authorities, US and European authorities negotiating with the dictatorship.

This long journey involved knowledge and practices on the move. Information about the new wonder drug travelled first on paper, in journals and correspondence. Paper, adapting to academic norms and crossing disciplinary and geographic boundaries, also moved through time. In addition to academic journals and private correspondence, information was also transmitted orally, a feature of human culture since ancient times and the most difficult route to follow historically. Oral transmission is undoubtedly the origin of some information presented by both newspapers and medical journals. As Darnton has suggested, there are many commonplace and mundane places where information is interchanged on a daily basis, including parks, bars, homes and infirmaries, and I suggest these were a major force in knowledge circulation concerning penicillin and its healing capacities. It is generally accepted that more than one person reads each issue of a weekly magazine or newspaper, while a medical journal in an academic or clinical setting is read by numerous colleagues.[7]

Broadcast news was an even more efficient method of information transmission during the period under study: people gathered to listen every day during wartime and the post-war confusion. It also had the advantage of being a private activity—enabling leftists to listen to BBC news when freedom of information, enjoyed by neighbouring democracies, was absent in dictatorship Spain—but the disadvantage of leaving little historical evidence beyond personal memories. Private recollections certainly exist relating to the history of penicillin, but I found no one that remembered hearing of the drug for the first time, nor a written report of that moment in time.

The reason for this review of communication channels in the early history of penicillin is that the circulation of knowledge on this new, efficient and non-toxic healing commodity was a key issue: knowledge created hope and demand. This was demand for a product, for more knowledge, and a desire to experience the new healing as a phenomenon embodied in the ampoules of penicillin. It created promises of cure, of a future free from the miseries of infection, both physical and symbolic. These expectations affected science as a social enterprise: the research laboratory was deemed the setting to produce cures, the factory would then manufacture, and the doctor would prescribe. An infection-free future was envisaged not only in Spain but across the globe. So promising was the action of penicillin that a utopian world free from disease seemed almost in reach.[8]

Reconstructing Journeys

No-Do reports, preserved at the state newsreel archives, provide evidence that public audiences knew of penicillin as early as June 1945, from a film on the use of penicillin by Allied troops.[9] Movies were highly popular at the time, and whether in a cinema or outdoors, the screening of No-Do news beforehand was compulsory. It was also in 1945 when the promise of the new drug was discussed at the Academia Nacional de Farmacia in Madrid and a number of books about penicillin were published in Spanish, by medical doctors such as José Álvarez Sierra, by the ophthalmologist Antonio Ros Sáez, and by the biologist and passionate champion of the drug, Florencio Bustinza. Others would soon follow, including a history of penicillin by the physician Alfonso Nadal Sauquet in 1946. Fleming himself authored two books, both published in 1948; a translation of *Penicillin: Its Practical Application* (1946) and a collection of his talks in Barcelona.[10] The end of the Second World War thus marked a moment from which to draw the trajectory of the dissemination of penicillin, as knowledge and as a commodity for infection cure. Venereal diseases were rarely mentioned, reports on the treatment of syphilis being restricted to medical journals.[11]

News on penicillin and its wondrous therapeutic activity circulated, and the treatment of syphilis by John Mahoney was reported in the Madrid daily *ABC* in October 1943. This was one of the few times when venereal diseases were mentioned in association with penicillin by the press. Another instance appeared some years later, in 1950, when national newspapers reproduced a long discourse by the minister of Gobernación (internal affairs) on government public health policy, in which he referred to prostitution and venereal diseases. Having briefly discussed the debate over whether prostitution should be prohibited or not, the minister stated his preference for 'control', noting the advantages for public health. The minister also stated that venereal infections had not been of great concern as sulfa-drugs and, subsequently, penicillin had so effectively diminished their impact.[12]

Six months after the 1943 Marine Hospital report came news of the first penicillin treatments in Spain, with Bustinza's involvement launching his career as spokesperson for the miraculous effects of the new antimicrobial drug, and for Fleming.[13] He became a familiar public face; his words disseminated in a way that has been preserved, evidence of the content of knowledge in circulation.

Newspapers, No-Do and broadcast news participated in the success of Alexander Fleming's visit to Spain in 1948. As his diaries reveal, he was recorded in every format available. In Chap. 2 I explored the symbolic value of penicillin in the misery of Spanish life during the 1940s, years of repression and hunger. The public adulation of Fleming belonged to a social life of symbols during this early decade of Franco's dictatorship, a social life of acknowledgements of the new state that had liberated Spain from 'red power'. Part of the population experienced relief; the other, devastation.

The captivating features of penicillin's healing abilities were represented in Fleming's lengthy visit. Applause followed the idol down the streets, to the bullfight, to cathedrals, museums, concerts and universities; gratitude was expressed with flowers at Las Rambles and in the words of monks at the Montserrat abbey. Recognition of the medical and political authorities was also exhibited, in every walk the Flemings took in Barcelona, Seville, Córdoba and Madrid, in private and public houses, in hospitals, medical academies and university halls, and in the shops where Fleming received gifts from thankful patients. Such public veneration of the drug was transferred to Fleming the discoverer. He felt the gratitude and adulation was excessive, beyond that which his achievements could justify, and he had a point: such fervour was embarrassing, and not only because it was overplayed, a social exaggeration of admiration for a dream expected to come true. This overacted homage was for the drug born when Fleming detected its activity in a Petri dish during the 1920s.

Fleming had arrived from a distant place, and distance played its part in creating the mythical status of penicillin.[14] Help would come from a place beyond the miseries of Spanish daily life; the otherness represented by penicillin would provide the cure. From far away, from other regimes of commodities and rights, of factories and hospitals, penicillin would eventually arrive in enough quantity and at a price 'for every pocket', as clinician Gregorio Marañón demanded in the prologue to Florencio Bustinza's first book. Penicillin would arrive, but in the meantime those people in the direst need would have to wait. Travel was expensive and passports not easy to obtain; only the very rich could travel in search of the products they wanted.

The majority of citizens waited for penicillin as they waited for nylons, coffee and tobacco: impatiently and anxiously. Fleming's role expanded from that of penicillin's representative to that of mediator: he was greeted as a supplier. But the substance was not materially travelling with

Fleming—there is no evidence he brought a single ampoule to Spain, neither did he visit any factory that could have manufactured the drug. Yet Fleming was a symbolic supplier. Time had passed since news of the miracle drug had been heard in Spain, and while time had not provided the drug, business eventually would. Fleming's visit was a kind of diplomacy, part of rebuilding the relationship between Franco's regime and the UK, a recovery based in trade: oranges and olive oil for penicillin.

While information about penicillin circulated, the first batches began to arrive, some through legal trade, others on the black market. One day the press would report the arrival of a legal batch; the next relate the tale of an engineer cured by penicillin of origins unknown to the administering doctor. Rationing circulated alongside knowledge and the drug itself. Penicillin had been strictly rationed in the USA during the Second World War, through early laboratory and clinical trials to mass production, reserved for soldiers at the front (see Chap. 4 and references therein). The smuggling of penicillin in the 1940s is not documented in the archives but remains in private and public recollections. Illegal trade endured during the 1950s—a period documented through the archival material of the Smuggling Courts—when the price of penicillin overseas had decreased. The evidence and public recollections suggest scarcity did not affect all social groups equally, but all levels participated in the black market and the early illegal circulation of the drug.[15]

During the decade of hunger and drought, penicillin joined the array of basic but scarce necessities; circulation involved the entrance of a commodity and knowledge associated with it to new socio-geographical domains. In the story narrated in this book geography indicates both mundane spaces—the street, public places, homes—and professional domains, dispensaries and clinics, and the bedsides of those in need. By penetrating these domains, penicillin participated in a culture of survival; the black market was a source of such survival for many, both buyers and sellers.[16] At the same time the new drug entered Franco's bureaucracy, as a controlled product with a set of norms to govern—or attempt to govern—its distribution. By participating in both the legal and illegal life of goods, penicillin became an agent in the history of both spaces. This reconstruction suggests it was an agent in the post-war history of Spain, the public perception of its healing capacities circulating in all the settings through which it moved. When penicillin was available, infections would be cured; this is what was believed, and for the lucky few of those early days, experienced.

Impediments to travel played a significant role.[17] These included the dictatorship's autarchic political economy; restrictions on the movement of people, money and goods; taxes, customs and the limited availability of foreign currency. The greatest concern for the militarised regime was defence, and industrial production was focused on transport and energy. Public health was not a priority; public order was. All these factors severely limited circulation. Political barriers prevented the entry of commodities—including books, which could be censored or banned completely—and hampered the creation of firms and industry. Such impediments operated in an international political climate of the post-war allied victory and played their part in the circulation of penicillin.

The trajectories for economic recovery taken by Allied countries became instrumental, leading to a political scenario composed of a restrictive national space amidst the Western European, democratic countries of the post-war era. The reorientation of Spanish policies was influenced by both national policies of neighbouring countries and international politics. Post-Second World War international diplomacy worked towards recovery, productivity, industrial and economic development. Foreign economic support included many benefits not intended for Franco's Spain. The Marshall Plan, the United Nations, UNESCO, OEEC and the World Health Organisation did try to thwart Franco's early strategies, but would later interact with the Spanish political regime. From the early 1950s the Cold War acted as a geo-strategy in the recovery of diplomatic relations with the USA, through the 1953 signing of a secret military agreement for US military bases on Spanish soil. US bases and American penicillin manufacturing licenses entered Spain together. The same year, the Concordat of 1953 with the Vatican was signed by Franco's minister of Foreign Affairs, thus establishing the two main pillars of the regime: the USA and the Catholic Church.[18] Medical care and the military were the bonds between Spain and the USA, while many religious orders provided care in Spanish hospitals, sanatoria and asylums.

Between the early isolation of Franco's Spain by international organisations and slow acceptance of the regime, the first decade and a half of the dictatorship had passed. Penicillin was starting to become freely available, and new penicillins had been developed. The drug was a tool for industrial development, through which agreements with US pharmaceutical manufacturers negotiated by Spanish firms intervened in the controlling practices and autarchy of Franco's policies; such international influences helped extend the dictatorship. Penicillin became a transnational agent in the international post-war recovery.[19] International organisations did not directly participate in establishing penicillin manufacturing plants in Spain, as they did in Poland, but in the early 1950s the World Health Organisation developed policies and provided funds for the control of venereal disease, used to purchase penicillin for Spanish sufferers as part of a cooperative programme.[20]

Penicillin finally travelled in large amounts. Imported in bulk, millions of units were bottled in Spain during 1950. The penicillin that first travelled officially was American, from Merck and Schenley, both US firms. With these huge amounts of penicillin, as part of the agreements, the methods for manufacturing the drug in two Spanish factories also travelled. This included the use of big fermentators operated by men and small, precise devices operated by women for weighing, bottling and labelling.

The genderised space of penicillin manufacturing assigned the delicate tasks of handling the pure drug in a pure environment to women, not only in Spain, and not only in penicillin factories, but in many manufacturing plants throughout Europe and the USA, before, during and after the war. By exploring circulation to bring additional agents to the history of penicillin—by giving agency to the act of travelling—these women workers are made visible. Their presence can be witnessed in a large number of photographs, not only of the CEPA plant but many antibiotic factories at that time, exhibited in publications and preserved in company archives. Either taken for granted or ignored, the participation of women—or 'the girls'—in the history of penicillin is emphasised here, and not only on the manufacturing line. Carmela Viejo, one of the technical heads of the factory, deserves a special mention; unfortunately I have been unable to find many biographical details, but her work among the staff contributes to the participation of women in penicillin manufacturing.

This suggests that a gendered order was attached to the manufacture of penicillin and many other commodities of the time. Recognising this retrieves more women in the history of penicillin, notably those who participated in the early research at Oxford, testing penicillin activity: Ethel Florey, Margaret Jennings and Jean Orr-Ewing were among the very early pioneers, signatories of the first papers on the new drug, who took part in the first clinical and bacteriological tests and wrote books on their experiences.[21] All studied medicine, Ethel Florey in Adelaide, Jennings and Orr-Ewing in Oxford. Mary Ethel Hayter Reed and Margaret Augusta Fremantle became obscured by the permanently reinforced figure of Howard Florey, who married his Australian colleague Ethel when she joined her fiancé in London, then Margaret the year after Ethel's death in 1966. It was from Florey's biography that biographical information about Margaret Jennings and Ethel Florey could be retrieved. Slightly later—penicillin travelled swiftly between the UK and USA—many other women were involved in testing penicillin in US laboratories during the Second World War; Gladys Hobby stands out, as a researcher, expert, author of the first high-quality, detailed reference book on the history of penicillin and director of a book series on chemotherapy, transformed under her direction into a journal, *Antimicrobial Agents and Chemotherapy*.[22]

Once the factory was operational and domestic penicillin manufacture had begun, the CEPA–Merck antibiotic screening program was set up in Madrid. Three young technicians were hired, one woman graduate in natural sciences, Sagrario Mochales, and two men, a chemist and a physician. My focus on Sagrario Mochales is based on her role in the detection of a new antimicrobial, originally identified, partially purified and tested by the CEPA–Merck group. The research programme took the form of a manufacturing line, generating microbiological cultures from which new antimicrobials could be isolated. The programme took an industrial shape, exhibiting an exchange of meanings and methods with the industrial production of a commodity that contradicted the linear model attributed to the relationship between research and industrial production. It reveals the influence of industry on research, a sequence also observed by Robert Bud in two US cancer research centres (see Chap. 5). This transfer of a method of production from the factory to the research laboratory participated in the culture of circulation.

The newly devised Sphero test, based on the production of spheroplasts, was designed by Eugene Dulaney at Merck Research Laboratories in Rahway. Details remained secret—nothing was published on the method until the 1990s—at a time when other pharmaceutical firms were also searching soil samples for new antibiotics, it was used by the Madrid screening programme for many years. Mochales detected a tiny antimicrobial activity in the cultures of a soil sample collected in Alicante; in vivo activity was demonstrated. From 1966, when detection was reported, through 1969, when the structure of the new compound was described by Merck chemists in Rahway, to 1970, when the procurement method was patented, Mochales attained increasing recognition as an industrial researcher. By that time she had been joined by Isabel Martin. Also a natural scientist, Isabel Martín participated in early research on fosfomycin by the CEPA screening team.

The screening laboratory became a research centre, Centro de Investigación Básica de España (CIBE), under Merck control in the 1980s, and Sagrario Mochales was appointed director. Its work on natural products continued until 2006, when what was by then a 50 year-old research centre closed.[23] A number of antibiotics had been identified during its history, three of which were marketed: fosfomycin, cephamycin and thienamycin. Cephamycin and thienamycin derivatives were marketed by Merck. Cephamycin C led to the development of cefoxitin, a semi-synthetic cephalosporin, widely used during the 1980s, and a derivative of thienamycin was marketed to treat bacterial infections in hospitals. Other drugs identified at CEPA's research laboratory included statins, to control atherosclerosis and reduce the risk of cardiovascular disease, and antifungal medications.[24] So penicillin inaugurated an era of antimicrobials in an increasingly ambitious culture of cure research and industry. This culture not only led to more antibiotics but also to other new pharmaceuticals that would intervene in contemporary expectations and experiences of healing.

Meanwhile, warnings about bacterial resistance to antibiotics were increasing. From the 1950s, the cure—a dream come true—was being at least partially eclipsed by concern about overprescription and misuse. The pathologist Mary Barber had presented evidence in 1947 of how resistant bacteria were disseminated in clinical wards, but it took a decade for relevant policy to be formulated. In the USA during the late 1950s, Max Finland and other experts in infectious disease would call for the establishment of controlled clinical trials and criticised the marketing practices of pharmaceutical firms every time a new antibiotic was put on the market. In Europe, Scandinavian countries took the lead in warning against practices that increased resistance and emphasised the need to regulate antibiotic use in farming (see Chap. 5). Penicillin, in all the domains it had entered, appeared to have created resistances in both the laboratory and bodies of patients. Doctors and scientists were aware of antibiotic resistance, and carrying out research on resistant bacteria from the late 1950s onward. Meanwhile, Spanish pharmaceutical firms were developing strategies to address the issue of contraband penicillin entering the market at lower prices than they were able to negotiate with health and industrial authorities.

It was in the 1980s that antibiotic consumption in Spain was fully evaluated. Teresa Pérez Gorricho collected a vast amount of data on antibiotic use in the clinic, veterinary medicine and animal fodder. Compulsory prescription and cautionary advice were implemented in a system increasingly aware of the need to combat overconsumption, protect patients and prevent infections. One of the most enduring achievements of the early democratic governments in Spain was establishing the legal basis for a national healthcare system: medical care as a universal right.[25] The importance of epidemiology as an instrument in health policy was recognised, and the area of expertise encouraged and financially supported. Wider studies on antibiotic consumption were undertaken.

Warnings against practices that could increase resistances circulated effectively, but the consumption of antibiotics did not decrease. Epidemiology and demography provided evidence of increases in both population and numbers of antibiotic treatments. Disease became largely synonymous with infectious disease, previously considered a temporary problem, soon to be consigned to history. In the 1990s the European Union created a surveillance system, distributing data on annual antibiotic consumption and bacterial resistance. This surveillance is a tool, a foundation from which to combat the increasing risk of resistant infections. Experts in Spain and the rest of Europe repeatedly publish warnings, advice and information. At the start of this millennium, 50 per cent of total antibiotic use was accounted for by human medicine, the other half by 'production of food for human consumption'.[26]

An Epistemology of Penicillin Circulation

In using the term circulation, I certainly do not mean circularity: this is not the story of a closed circle traversed by the drug. Circulation refers to traffic, the movement of people, methods, materials and skills. As the first 'antibiotic', penicillin occupied a diversity of domains, undergoing journeys from promise to materiality, from dreams to cures, and constructing its identity as a transnational object in the healing cultures of the mid- to late twentieth century.

Although penicillin has produced resistances in a number of widely distributed strains, its derivatives remain the most consumed group of antibiotics, while antimicrobial activity prevails as one of the most enduring developments produced by twentieth-century biomedicine. Antibiotic resistance may well be regarded a part of this culture: mutated bacteria and non-mutated bacteria travel equally freely, as freely as infections. Contemporary travel—increasingly cheaper, faster and easier to access, and therefore, arguably overconsumed—carries with it both antibiotic treatment and resistance.

The new antibiotic detected in Madrid, fosfomycin, circulated from the research laboratory, through the specially designed pilot plant, to a new factory for chemical manufacture; synthesis of the drug proved more efficient than the fermentation that produced penicillin derivatives and streptomycin. As part of the paper technologies generated by the patent system—protecting the procurement method, not the drug—such methods also travelled. Clinical trials were established in the USA and Spain.

Rush by Akash Kapur is an inspiring story of a big road meeting a small village.[27] As the road is being constructed, new ambitions are created. The road itself is a way of reaching other places, cultures and knowledge; it provides a path to the irresistible attraction of the faraway. The trajectory taken to reach such places—in the case of penicillin, infection cure and the entire system of research and industry that produced the early batches—became an objective in itself, alongside ambitions to make the drug permanently available. This is what circulation aspires to speak about: how the path itself becomes the aim. As penicillin crossed the ocean, a process of selection, transfer and transformation took place. Products were in motion; expeditions of people, tools and their associated knowledge were created and re-created in these transits embedded in a history of communication.[28]

This culture spread through the written word and the experience of infection cure. All along the journeys information was exchanged, action at a distance was enabled and a symbolic culture was shared.[29] The vehicles used for transportation took part, as did customs and the Spanish government's control of foreign currency; back and forth across national borders. Expeditionary people and objects created, maintained and extended expert communities. The particular trajectories of the many objects that participated in the circulation of penicillin in Spain, and the connected subjects and ideas, are at the core of this historical narrative, which emphasises transportation, diffusion, appropriation and learning. Ideas and people, places and journeys, from Spain and beyond are retrieved.

Times and spaces provide this reconstruction of penicillin's history with a set of analytical categories. Gender and political history interacted with the commodity penicillin became, both in itself and through the generation of other commodities. In the wide zone of exchanges where there was a continent—Europe—and a geographical and political culture—the West—located in the post-Second World War era, both the material and the symbolic circulated. By connecting with habits and beliefs, instruments and knowledge were sanctioned, as were gendered distributions of tasks and authority. The set of knowledge and practices attached to penicillin carried a system of authority. Some people had knowledge about penicillin—its composition and origin, its medical use—others became trained and skilled, ready to produce further knowledge and materials. As Carlo Belfanti has stated, 'the dynamic circulation of technical know-how' can be seen when focusing 'closely at the various national realities'; 'the diffusion of technical knowledge via the movement of peoples' gains agency as it takes 'many different shapes, varying with period and place'.[30]

The movement of penicillin took place within the West and the East, throughout metropolitan geography. These movements were smooth for the most part, through shared cultures; a culture of travels as sources of knowledge and practices. However, travel implies asymmetry: movements took place in both directions but the impacts were not equal, constrained by the geographic distribution of authority, technical capabilities and wealth. These travels helped stabilise the authority of those who created the object: the UK and the USA were assured their place in the history of the healing cultures inaugurated by penicillin—following the early toxicity of sulfa-drugs—and their success was based on the wide distribution, expectation, acceptance and adoption of the biomedical knowledge carried in each box of ampoules. Between laboratories and pharmacies were the bedside, medical authority and the factory. Penicillin took on a diversity of identitics as it was transferred across worlds. It linked sites of production with sites of consumption through medical authority.

By understanding the culture that surrounded circulation, the transformation in asymmetrical power relationships, and the creation of authority based on technologies and distribution policies, the history of penicillin widened its scope, geographically and conceptually. The details of its everyday life reveal the scientific concerns of both those who produced and dispatched, and those who received the new drug.

For producers, a strategy of protecting rights and tools carried a powerful baggage of promises of authority. Authority as power also circulated.[31] This included the conventions surrounding women's work, dictatorship policies, the strategies of Allied recovery and US policies to assist, diplomacy and the research laboratory, and autarchy and its failings: such is the impossibility of manufacturing any commodity in isolation. As Soraya Budia has stated, 'financial investment and the collaborations involved finally managed to generate a social reality for the medical applications'.[32]

This has not been the biography of a drug in Spain, no matter how marvellous its effects may have been. It is a narration on the multitude of places and times that penicillin lived in, and moved through, as if acquiring a life of its own, as Ian Hacking has described the process scientific objects undergo. Beyond the trend towards global narratives or any apparent homogenisation through standardisation, nations remain relevant within the history of antibiotics.[33]

The drug was born in a Petri dish between London and Oxford, and as many other epistemic things produced in the twentieth-century laboratory, it was both cultured and participated in creating a culture, transiting between microbiology and society. And at each stop, penicillin left a mark that can be traced by the historian seeking an explanation of the lifestyle it created. Penicillin has not been regarded as an isolated object in this reconstruction, but one embedded in the laboratory practices of microbiology, a delicate product to be manufactured, contained, transported and anxiously awaited.

Penicillin, as a scientific and medical object of twentieth-century history, is useful for revealing the many linked practices in different social and cultural domains. Clinicians, researchers, patients and health authorities, as well as fictional and non-fictional recollections, shaped the cultures of antibiotics. Prescription regulations for the new drugs were part of this history, connected to the emergence of welfare state policies in a number of European countries. Agents, however, acted locally, and as Alexei Kojevnikov has described in relation to Soviet totalitarianism, local secrecy and lack of public discussion took effect.[34] In the silent space it inhabits, secrecy, as Susan Lindee has described, involves struggles and strategies, as well as policies and social interactions.[35]

Gender as an agent played a role in the research laboratory and the manufacturing plant. These two gendered spaces of production reveal the similarities between the production of scientific and medical knowledge and that of industrial objects. Knowledge and objects are both commodities manufactured and transported in contemporary societies. Gender remains encoded in both, with men controlling management of the workplace and the distribution of recognition. In both, women's access has initiated shifts in cultural and social norms, modifying the symbolic values attached to particular skills in the workplace and in society at large. The reconstruction of penicillin's travels thus becomes useful for a historical account aspiring to widen the scope of the role played by the drug, and by medicine in general, to better understand its success and its place in contemporary cultures of healing and consumption.

Notes

1. Inspired by Robert Kohler's term 'breeding reactor', which he applies to the breed *Drosophila melanogaster* as an experimental system in Thomas H. Morgan's laboratory. Robert E. Kohler, *Lords of the Fly: Drosophila Genetics and the Experimental Life* (Chicago: University of Chicago Press, 1994).
2. Hanna Landecker uses this metaphor when talking about tissue culture. Hannah Landecker, *Culturing Life: How Cells Became Technologies* (Cambdrige, MA: Harvard University Press, 2007).
3. Ana Romero de Pablos, 'Regulation and the circulation of knowledge: Penicillin patents in Spain', *Dynamis* 31(2) (2011): 363–83.
4. In mid-1943 the US government was paid 20 dollars a dose. By the end of the war the price had decreased to 6.5 cents. See Robert Bud, *Penicillin: Triumph and Tragedy* (Oxford: Oxford University Press, 2008), 53.
5. Maria Rentetzi, *Trafficking Materials and Gendered Experimental Practices: Radium Research in Early 20th Century Vienna* (New York: Columbia University Press, 2008).
6. Robert Darnton, 'What is the history of books?' *Daedalus* 111 (1982): 65–83. See also his 'An early information society: News and the media in eighteenth-century Paris', *American Historical Review* 105 (2000): 1–35.
7. I do not have the satisfying evidence James Secord has recorded; see his *Victorian Sensation: The Extraordinary Publication, Reception, and Secret Authorship of Vestiges of the Natural History of Creation* (Chicago: University of Chicago Press, 2003).
8. For a review of the argument from the history of antibiotics in the USA see Scott Podolsky, *The Antibiotic Era: Reform, Resistance, and the*

Pursuit of Rational Therapeutics (Baltimore: Johns Hopkins University Press, 2014).

9. No-Do Noticiario Nº 130ª, June 25, 1945.
10. José Álvarez Sierra, *Lo que cura la Penicilina. Presente y porvenir de una droga mágica* (Madrid: Más allá, 1945); Alfonso Nadal Sauquet, *Historia de la penicilina* (Barcelona: Lauro, 1946); Florencio Bustinza, *Los antibióticos antimicrobianos y la penicilina: De Pasteur a Fleming* (Madrid: Plus Ultra, 1945); Alexander Fleming, *Penicilina: su aplicación y práctica* (Madrid: Saturnino Calleja, 1948).
11. Xavier Vilanova, 'El tratamiento penicilínico de la sífilis', *Revista Clínica Española* 58 (1955): 307–20.
12. *ABC*, July 13, 1950, 10; *La Vanguardia*, July 14, 1950, 4.
13. 'Un nuevo remedio contra la infección luética', *ABC*, October 17, 1943, 35.
14. Paola Bertucci, 'The in/visible woman: Mariangela Ardinghelli and the circulation of knowledge between Paris and Naples in the eighteenth century', *Isis* 104 (2013): 226–49.
15. Óscar Rodríguez Barreira, 'Cambalaches: hambre, moralidad popular y mercados negros de guerra y postguerra', *Historia Social* 77 (2013): 149–74, on 162.
16. Rodríguez Barreira, 'Cambalaches'.
17. I am grateful to John Krige, who asked me about the impediments to circulation at the Dark Matters Conference in Barcelona, 2013.
18. This also applied to radioisotopes during the same period. See María Jesús Santesmases, 'From prophylaxis to atomic cocktail: Circulation of radioiodine', *Dynamis* 29 (2009): 337–63.
19. See Simone Turchetti, Nestor Herran and Soraya Boudia, 'Introduction: Have we ever been "transnational"? Towards a history of science across and beyond borders', *The British Journal for the History of Science* 45 (2012): 319–36; Edna Suárez-Díaz, Gisela Mateos and Ana Barahona, 'Across borders: Science and technology during the Cold War: An introduction', *Dynamis* 35 (2015): 271–8. For the situation in Poland see Sławomir Łotysz, 'Controlling the production and distribution of drugs in communist Poland', *Medicina nei Secoli: Arte e Scienza* 26 (2014): 519–56.
20. Ramón Castejón Bolea and Esteban Rodríguez Ocaña, 'El control de las enfermedades venéreas en España durante la década de 1950: los inicios del proyecto España 8 de la OMS', paper presented at the XVII Congreso de la Sociedad Española de Historia de la Medicina, San Feliu de Guixols, June 15–17, 2017.
21. On Orr-Ewing see the obituary by A. D. Gardner, 'Jean Orr-Ewing. Born 28th April 1897. Died 17th November 1944', *The Journal of Pathology* 58 (1946): 149.

22. Gladys L. Hobby, 'Antimicrobial agents and chemotherapy', *ASM News* 40 (1974): 327–9; Gladys L. Hobby, *Penicillin: Meeting the Challenge* (New Haven: Yale University Press, 1985).
23. A new centre was created later in 2008 at the University of Granada, which took advantage of CIBE's experience and techniques with the support of the regional government of Andalusia. Limited information is available at www.granadaeconomica.es/06/2013/fundacion-medina-centro-excelencia-investigacion-medicamentos (last accessed July 21, 2017).
24. Fernando Peláez, 'Introduction: 50 years of CIBE', in *CIBE: 50 años de investigación* (Madrid: Merck Sharp and Dhome España, 2005), 23–33, on 24–5. See also Fernando Peláez, 'The historical delivery of antibiotics from microbial natural products—Can history repeat?', *Biochemical Pharmacology* 71 (2006): 981–90.
25. *Boletín Oficial del Estado*, 29 de marzo de 1986: Ley 14/1986, de 25 de abril, General de Sanidad.
26. Agencia de Evaluación de Tecnologías Sanitarias, Ministerio de Sanidad y Consumo, 'Informe sobre resistencia microbiana. ¿qué hacer?' *Medicina Clínica* 106 (1995): 267–79; José Campos and Fernando Baquero, 'Resistencia a antibióticos: ¿qué hacer ahora?', *Medicina Clínica* 119 (2002): 656–8.
27. Akash Kapur, 'Rush', *The New Yorker*, October 14, 2013.
28. James Secord, 'Knowledge in transit', *Isis* 95(4) (2004): 654–72.
29. Rima D. Apple, Gregory J. Downey and Stephen L. Vaughan, eds., *Science in Print: Essays on the History of Science and the Culture of Print* (Madison: University of Wisconsin Press, 2012).
30. Carlo Belfanti, 'Guilds, patents, and the circulation of technical knowledge: Northern Italy during the early modern age', *Technology and Culture* 45(3) (2004): 569–89.
31. Maria Rentetzi, *Trafficking Materials and Gendered Experimental Practices: Radium Research in Early 20th Century Vienna* (New York: Columbia University Press, 2008).
32. Soraya Boudia, 'Radioisotopes "economy of promises": On the limits of biomedicine in public legitimization of nuclear activities', *Dynamis* 29 (2009): 241–59, on 244.
33. Andrea Tone and Elizabeth Siegel Watkins, eds., *Medicating Modern America: Prescription Drugs in History* (New York: New York University Press, 2007); Arthur Daemmrich, *Pharmacopolitics: Drug Regulation in the United States and Germany* (Chapel Hill and London: University of Carolina Press, 2004).

34. Alexei Kojevnikov, 'The making of the Soviet bomb and the shaping of Cold War science', In *Reappraising Oppenheimer: Centennial Studies and Reflections*, eds. C. Carson and D. A. Hollinger (Berkeley: Office for History of Science and Technology, 2005), 129–45.
35. M. Susan Lindee, 'Keeping secrets, learning to lie', paper presented at Dark Matters: Contents and Discontents of Cold War Science International Conference, Barcelona, May 31–June 2, 2013.

CHAPTER 9

Final Reflections

Penicillin, Infections and Resistances: Past and Present

The long, impressive history of penicillin was made possible by the detection of a particular activity in a culturing media; a round circle in paper would eventually reveal that activity as one microbe killing another. A substance produced by a microbial was the poison, rather than the microbial itself. And so the antibiotic era became molecular, a trajectory that became one of the defining features of the last half of the twentieth century. Here was a product, a substance with a chemical composition and, following X-ray studies by Dorothy Hopkins and Barbara Low in 1945, a three-dimensional structure.

But the molecule was not the focus, the aim or even the origin of penicillin's history; that was a powder extracted from a mould. Penicillin, like other scientific and technological products of twentieth-century research and innovation, provided a cure and created new phenomena, not all to be welcomed. The fact that microbes not only produce antimicrobials but also a counteracting enzyme, penicillinase, was an early threat. Antibiotics created mutant microorganisms through what Watanabe called the R-factor (for resistance), later named plasmids; these cytoplasmic units became one of the main circulators of resistance.

Resistances have followed the same patterns as infections. Blighted by pandemics for centuries, humanity's main defences have been confinement and quarantine. Time has created a genealogy of infections,

M. J. Santesmases, *The Circulation of Penicillin in Spain*,
Medicine and Biomedical Sciences in Modern History,
https://doi.org/10.1007/978-3-319-69718-5_9

a history of pandemics and the origins of epidemiology. It is by drawing this genealogy that resistance to antibiotics becomes embedded in the history of infections, even if this particular case appeared during the second half of the twentieth century in Europe, Japan and America, through the combination of bacterial behaviour with antibiotics. Substances produced by other microbes have combated those bacteria many times before producing a reaction, and these reactions have experienced many journeys, between wards, hospitals and countries. The phenomenon itself also produced a reaction, an attempt to prevent overconsumption. In search of explanations for resistance, studies of antibiotic use—on expenses, success and failures—have been systematically collected.

In Spain, legislation has been amended alongside changes in regime. Between the dictatorship's *Bases de Sanidad Nacional* in 1944, to *Ley General de Sanidad*, approved by democratic congress in 1986, over forty years have passed. This has been the period studied and presented in this book. The Francoist law kept infections under the jurisdiction of a repressive government, which viewed them as a matter of civilian and political order. Any unusual disorder had to be declared and the sick isolated as if politically suspicious; a further punishment. The 1986 law was intended to protect both the ailing individual and the citizenry as a whole, to prevent the spread of infections and provide care. The political changes in Spain during this period became embedded in international dialogue, in negotiations for the right to access primary care and the emergence of epidemiology as a scientific foundation of health policy. The advancement of epidemiology has maintained compulsory declaration, as a public health matter: the patient is no longer under suspicion. The availability of penicillin increased dramatically over this long period, beyond early expectations during the Years of Hunger. Overuse in medicine and agriculture has now created new microbial and medical ecologies: new dangers.

Unknown infections and pandemics have appeared, AIDS being the most dramatic and challenging. An entire system based on heterosexual social order was challenged while viruses were brought into the medical drug culture of healing. Viruses, whose biological identity was defined as somewhere between a set of molecules and a living being, have resisted both classification and antibiotic action.

The politics and power of the dictatorship did not occupy all spaces, but had a hand in most. The exiles within, those who resisted in a modest way, participated in making Spanish society ready to take advantage of the international networks of knowledge and practices first accessible during the 1960s. They were not alone, as political authorities became practised in playing in the troubled waters—for Franco at least—of a democratic European continent and its international institutions. Alongside the circulation of penicillin and later antibiotics, a culture of research and an expectation of a free world—free from infections and the arbitrary powers of a dictatorship—also emerged. Freedom occupied the political economy of the country before any political right could be exercised by the people, however, and penicillin joined this space of economic liberalism without civil rights. Research and expert communities have increasingly legitimised hierarchies, producing the so-called 'knowledge society' in which we live.

Through the events that have formed recent contemporary history, penicillin has retained its agency, participating in the miseries and wonders of everyday life, healthcare and cure, science and medicine, and in the political events of its time. Penicillin's coexistence with the resistance it created has also endured: activity and resistance remain, an incongruous partnership that has produced the new microorganistic ecologies found in healthcare settings today.

Bibliography

Abraham, E., E. Chain, C. M. Fletcher, A. D. Gardner, N. G. Heatley, Margaret A. Jennings, and H. W. Florey, 'Further observations on penicillin', *The Lancet* 238 (1941): 177–89.

Adams, David P., 'Wartime bureaucracy and penicillin allocation: The committee on chemotherapy and other agents', *Journal of the History of Medicine and Allied Sciences* 44 (1989): 196–217.

Adams, David P., *The Greatest Good to the Greatest Number: Penicillin Rationing on the American Home Front, 1940–1945* (New York: Peter Lang, 1991).

Agencia de Evaluación de Tecnologías Sanitarias, Ministerio de Sanidad y Consumo, 'Informe sobre resistencia microbiana. ¿Qué hacer?', *Medicina Clínica* 106 (1995): 267–79.

Álvarez-Sierra, José, *Lo que cura la Penicilina* (Madrid: Más allá, 1945).

Anderson, Benedict, *Imagined Communities: Reflections on the Origin and Spread of Nationalism* (London: Verso Books, 2006).

Anonymous, *Revisión de los conocimientos actuales sobre la Penicilina* (Chicago: Abbot International Laboratories Co., 1946).

Apple, Rima D., Gregory J. Downey, and Stephen L. Vaughan, eds., *Science in Print: Essays on the History of Science and the Culture of Print* (Madison: University of Wisconsin Press, 2012).

Artola Blanco, Miguel, *El fin de la clase ociosa: de Romanones al estraperlo, 1900–1950* (Alianza Editorial, 2015).

Ballabriga Aguado, A., 'Algunas consideraciones acerca de la penicilina y otras micoínas,' *Medicina Clínica* IV 2 (1945): 152–6.

Baltz, R. H., 'Antibiotic discovery from actinomycetes: Will a renaissance follow the decline and fall?,' *SIM News* 55 (2005): 186–96.

M. J. Santesmases, *The Circulation of Penicillin in Spain*, Medicine and Biomedical Sciences in Modern History,
https://doi.org/10.1007/978-3-319-69718-5

Baquero, Fernando, Manuel Martínez Ferrer, and Ana Carvajal, 'Ecología de la transmisión de resistencias plasmídicas a los antobióticos en el intestino humano', in *Aspectos actuales en la transmisión de la información genética de microorganismos*, eds. Antonio Portolés and Fernando Baquero (Madrid: Sociedad Española de Microbiología, 1975), 267–99.

Barber, Mary, 'Coagulase positive staphylococci resistant to penicillin,' *Journal of Pathology* 59 (1947): 373–84.

Barber, Mary, 'Staphylococcal infection due to penicillin-resistant strains,' *British Medical Journal* 2 (1947): 863–865.

Barciela, Carlos, 'Crecimiento y cambio en la agricultura española desde la Guerra Civil', in *La economía española en el siglo XX. Una perspectiva histórica*, eds. Jordi Nadal, Albert Carreras and Carles Sudriá (Barcelona: Ariel, 1987), 258–79.

Barnett, Jack A., Paul M. Southern Jr., James P. Luby, and Jay P. Sanford, 'Eficacy of phosphonomycin in treatment of urinary-track infections', *Antimicrobial Agents and Chemotherapy* 9 (1969): 349–51.

Barona, J. L., 'Un exili de transició. Científics republicans a Gran Bretanya,' *Mètode* 61 (2009): 72–9.

Barona, J. L., and E. Perdiguero-Gil, 'Health and the war. Changing schemes and health conditions during the Spanish Civil War', *Dynamis* 28 (2008): 103–26.

Belfanti, Carlo, 'Guilds, patents, and the circulation of technical knowledge: Northern Italy during the early modern age', *Technology and Culture* 45(3) (2004): 569–89.

Berg, Maxine, *The Age of Manufactures, 1700–1820: Industry, Innovation and Work in Britain* (London: Routledge, 2005).

Bertucci, Paola, 'The in/visible woman: Mariangela Ardinghelli and the circulation of knowledge between Paris and Naples in the eighteenth century', *Isis* 104 (2013): 226–49.

Blume, Stuart, Christine Holmberg, and Paul Greenough, eds., *The Politics of Vaccination. A Global History* (Manchester: Manchester University Press, 2017).

Bonah, Christian, and Anne Rasmussen, eds., *Histoire et medicament aux 19e et 20e siècles* (Paris: Gliphe, 2005).

Bonah, Christian, Christophe Masutti, Anne Rasmussen and Jonathan Simon, eds., *Harmonizing Drugs: Standards in Twentieth-century Pharmaceutical History* (Paris: Glyphe, 2009).

Borderías Mondéjar, Cristina, *Entre líneas: trabajo e identidad femenina en la España contemporánea: la Compañía Telefónica, 1924–1980* (Barcelona: Icaria, 1993).

Borowy, Iris, 'Global health and development: Conceptualizing health between economic growth and environmental sustainability', *Journal of the History of Medicine and Allied Sciences* 68 (2013): 451–85.

Boudia, Soraya, 'Radioisotopes "economy of promises": On the limits of biomedicine in public legitimization of nuclear activities', *Dynamis* 29 (2009): 241–59.

Box Varela, Zira, *España, año cero: la construcción simbólica del franquismo* (Madrid: Alianza Editorial, 2010).

Bravo Ortega, A., R. Pueyo García, and A. Martínez Martínez, 'Nuestra experiencia en el tratamiento de la meningitis tuberculosa', *Revista Clínica Española* 51 (1953): 397–411.

Brenner, S., F. A. Dark, P. Gerhardt, M. H. Jeynes, O. Kandler, E. Kellenberg, Emmy Klieneberger-Nobel, K. McQuillen, M. Rubio-Huertos, M. R. J. Salton, R. E. Strange, J. Tomsik and C. Weibull, 'Bacterial protoplasts', *Nature* 181 (1958): 1713–15.

Brown, Kevin, *Penicillin Man: Alexander Fleming and the Antibiotic Revolution* (Gloucestershire: Sutton, 2004).

Bud, Robert F., 'Strategy in American cancer research after World War II: A case study', *Social Studies of Science* 8 (1978): 425–59.

Bud, Robert, 'Antibiotics. From germophobia to the carefree life and back again: The lifecycle of the antibiotic brand', in *Medicating Modern America: Prescription Drugs in History*, eds. Andrea Tone and Elizabeth S. Watkins (New York: NewYork University Press, 2007), 17–41.

Bud, Robert, *Penicillin: Triumph and Tragedy* (Oxford: Oxford University Press, 2008).

Bud, Robert, *The Uses of Life* (Cambridge: Cambridge University Press, 1993).

Bugallal, Isabel, 'Walter Starkie, el Borrow irlandés', at www.laopinioncoruna.es/contraportada/2010/12/14/walter-starkie-borrow-irlandes/447969.html.

Burns, Marlene, 'Scientific research in the Second World War. The case for Bacinol, Dutch penicillin', in *Scientific Research in World War II. What Scientists Did in the War*, eds. Ad Maas and Hans Hooijmaijers (London and NewYork: Routledge, 2009), 44–61.

Burris, R. H., and E. H. Newcomb, 'Kenneth Bryan Raper', *Biographical Memoirs of the National Academy of Sciences* 60 (1991): 251–72.

Bustinza, Florencio, *Diez años de amistad con Sir Alexander Fleming* (Madrid: M.A.S., 1961).

Cabré, Anna, 'Cuatro aproximaciones explicativas a las tendencias de nupcialidad y fecundidad', in *La constitución familiar en España*, eds. A. Cabré (Madrid: Fundación BBVA, 2007).

Cabré i Pairet, Montserrat, 'Women or healers? Household practices and the categories of health care in late medieval Iberia', *Bulletin of the History of Medicine* 82 (2008): 18–51.

Cabrera, Mercedes, *Juan March 1880–1962* (Madrid: Marcial Pons, 2011).

Camí Morell, Jordi, M. A. Zulueta, M. Teresa Fernández, María Bordons and Isabel Gómez Caridad, 'Producción científica española en biomedicina y ciencias de la salud durante el periodo 1990–1993 (Science Citation Index y Social Science Citation Index) y comparación con el periodo 1986–1989', *Medicina Clínica* 109 (1997): 481–96.

Campos, José, and Fernando Baquero, 'Resistencia a antibióticos: ¿qué hacer ahora?' *Medicina Clínica* 119 (2002): 656–8.

Camprubí, Lino, 'One grain, one nation: Rice genetics and the corporate state in early Francoist Spain (1939–1952)', *Historical Studies in the Natural Sciences* 40 (2010): 499–531.

Camprubí, Lino, *Engineers and the Making of the Francoist Regime* (Cambridge, MA: MIT Press, 2014).

Camprubí, Lino, *Los ingenieros de Franco: Ciencia, catolicismo y Guerra Fría en el Estado franquista* (Barcelona: Crítica, 2017).

Capocci, Mauro, 'A chain is gonna come: Building a penicillin production plant in post-war Italy', *Dynamis* 31 (2011): 343–62.

Carlsson, Bo, 'The development and use of machine tools in historical perspective', *Journal of Economic Behaviour and Organisation* 51 (1984): 91–114.

Cassidy, Patrick J., and Frederick M. Kahan, 'A stable enzyme phosphoenolpyruvate intermediate in the synthesis of Uridine-5'-diphospho-N-actey l-2-amino-2-deoxyglucose 3-0-Enolpyruvyl Ether', *Biochemistry* 12 (1973): 1364–75.

Castejón Bolea, Ramón, 'Access to penicillin for the treatment of syphilis in Spain in the 1950s: The role of the international agencies', *European Social Science History Conference*, Valencia, March 30, 2016.

Castejón Bolea, Ramón, *Moral sexual y enfermedad: la medicina española frente al peligro venéreo (1868–1936)* (Granada-Alicante: Universidad de Granada-Instituto de Cultura Juan Gil-Albert, 2001).

Castejón Bolea, Ramón, and Esteban Rodríguez Ocaña, 'El control de las enfermedades venéreas en España durante la década de 1950: los inicios del Proyecto España 8 de la OMS', *Al servicio de la salud: la historia de la medicina ante los retos del siglo XXI. Congreso de la Sociedad Española de Historia de la Medicina*, San Feliu de Guixols, June 2017, A. Zarzoso and J. Arrizabalaga, eds. (San Feliu de Guixols: SEHM, 2007) 101–108.

Catalán, Jordi, *La economía española y la segunda guerra mundial* (Barcelona: Ariel, 1995).

Cazorla Sánchez, Antonio, *Fear and Progress: Ordinary Lives in Franco's Spain, 1939–1975* (Maldon, MA: Wiley, 2009).

Cazorla Sánchez, Antonio, *Las políticas de la victoria: la consolidación del Nuevo Estado franquista, 1938–1953* (Madrid: Marcial Pons Historia, 2000).

Cebrián, Mar, 'Technological Imitation and Economic Growth During the Golden Age in Spain: 1959–1973' (PhD diss., Florence: European University Institute, Florence, 2004).

Cenarro Lagunas, Ángela, *La sonrisa de Falange: Auxilio Social en la guerra civil y en la posguerra* (Barcelona: Crítica, 2006).

Cenarro Lagunas, Ángela 'Los días de la "Nueva España": entre la "revolución nacional" y el peso de la tradición', *Ayer* 51 (2003): 115–34.

Chain, Ernst, H. W. Florey, A. D. Gardner, N. G. Heatley, Margaret A. Jennings, Jean Orr-Ewing, and A. G. Sanders, 'Penicillin as a chemotherapeutic agent', *The Lancet* 236 (1940): 226–8.

Chamorro, Jesús, *Recuerdos de una vida* (Oñate: Diletante, 2002).

Chapman, Stanley D., 'The textile factory before Arkwright: A typology of factory development', *The Business History Review* 48 (1974): 451–78.

Christiansen, Thomas, 'Crime and punishment in the black market in Spain, 1940–1950: An analysis of the Fiscalía de Tasas', *Ager* 4 (2005): 63–99.

Christinesen, B. G., W. J. Leanza, T. R. Beattie, A. A. Patchett, B. H. Arison, Robert E. Ormond, Frederick A. Kuehl, Jr., G. Albers-Schonberg, and O. Jardetzky, 'Phosphonomycin: Structure and synthesis', *Science* 160 (1969): 123–5.

Cifuentes Delatte, L., 'Tratamiento actual, médico y quirúrgico, de la tuberculosis del aparato urinario', *Revista Clínica Española* 53 (1954): 281–303.

Cipolla, Carlo Maria, *Faith, Reason, and the Plague in Seventeenth-Century Tuscany* (New York: W.W. Norton & Company, 1981).

Clark, H., N. K. Brown, J. F. Wallace, and M. Turck, 'Evaluation of phosphomycin, a new cell wall-active antibiotic', *Antimicrobial Agents and Chemotherapy* 9 (1969): 338–42.

Comroe, Julius H., 'Pay dirt: The story of streptomycin part II', *American Review of Respiratory Disease* 117 (1978): 957–68.

Condrau, Flurin, and Robert G. W. Kirk, 'Negotiating hospital infections: The debate between ecological balance and eradication strategies in British hospitals, 1947–1969', *Dynamis* 31 (2011): 385–405.

Coyette, Jacques, Jean-Marie Frère and Peter Reynolds, 'Jean-Marie Ghuysen', *Molecular Microbiology* 57 (2005): 871 3.

Cozzoli, Daniele, 'Penicillin and the European response to post-war American hegemony: The case of Leo-penicillin', *History and Technology* 30 (2014): 83–103.

Creager, Angela N., 'Adaptation or selection? Old issues and new stakes in the postwar debates over bacterial drug resistance', *Studies in History and Philosophy of Biological and Biomedical Sciences* 38 (2007): 159–90.

Crous Serdá, Antonio, 'Septicemia estafilocócica tratada y curada con penicilina', *Medicina Clínica* 4(5) (1945): 415–19.

Cunningham, Andrew, and Perry Williams, *The Laboratory Revolution in Medicine* (Cambridge: Cambridge University Press, 2002).

Daemmrich, Arthur 'Synthesis by microbes or chemists? Pharmaceutical research and manufacturing in the antibiotic era', *History and Technology* 25 (2009): 237–56.

Daemmrich, Arthur, *Pharmacopolitics: Drug Regulation in the United States and Germany* (Chapel Hill and London: University of Carolina Press, 2004).

Darnton, Robert, 'An Early Information Society: News and the Media in Eighteenth-Century Paris', *American Historical Review* 105 (2000): 1–35.

Darnton, Robert, 'What is the history of books?' *Daedalus* 111 (1982): 65–83.

de Chadarevian, Soraya, and Harmke Kamminga, eds., *Molecularizing Biology and Medicine: New Practices and Alliances, 1910s–1970s* (Amsterdam: Harwood, 1998).

de Grazia, Victoria, *Irresistible Empire: America's Advance Through Twentieth-Century Europe* (Cambridge, MA: Harvard University Press, 2005).

de Miguel, Amando, *España oculta: la economía sumergida* (Madrid: Espasa Calpe, 1988).

del Arco Blanco, Miguel Ángel, 'El estraperlo: pieza clave en la estabilización del régimen franquista', *Historia del Presente* 15 (2010): 65–78.

del Arco Blanco, Miguel Ángel, 'Hunger and the consolidation of the Francoist Regime (1939–1951)', *European History Quarterly* 40 (2010): 458–83.

del Arco Blanco, Miguel Ángel, *Las alas del Ave Fénix: La política agraria del primer franquismo (1936–1959)* (Granada: Comares, 2005).

del Cura, Isabel, and Rafael Huertas, *Alimentación y enfermedad en tiempos de hambre. España, 1937–1947* (Madrid: CSIC, 2007).

Delaunay, Jean-Marc, 'La liquidation des avoirs allemands en Espagne (1945–1961)', in *España, Francia y la Comunidad Europea*, eds. Jean-Pierre Étienvre and José Ramón Urquijo (Madrid: CSIC-Casa de Velázquez, 1989), 219–45.

Delgado, Lorenzo, 'Las relaciones culturales entre España y Estados Unidos, de la Guerra Mundial a los Pactos de 1953', *Cuadernos de historia contemporánea* 25 (2003): 35–59.

Delgado, Margarita, 'Cambios recientes en el proceso de formación de la familia', *Revista Española de Investigaciones Sociológicas* 64 (1993): 125–54.

Dulaney, Eugene L., A. H. Larsen, and Edward O. Stapley, 'A note on the isolation of microorganisms from natural sources', *Mycologia* 47 (1955): 420–2.

Esteve, A., and A. Oriol Anguera, 'Aportaciones experimentales a la farmacognosia penicilínica', *Medicina Clínica* 3(2) (1944): 116–28.

Fleming, Alexander, *Conferencias magistrales* (Barcelona and Madrid: Publicaciones del Hospital Municipal de Infecciosos-Diana Artes gráficas, 1948).

Fleming, Alexander, *Penicilina: su aplicación y práctica* (Madrid: Saturnino Calleja, 1948).

Florey, Howard W., Ernst B. Chain, Norman G. Heatley, Margaret A. Jennings, A. G. Sanders, Edward P. Abraham, and M. Ethel Florey, *Antibiotics: A Survey of Penicillin, Streptomycin, and Other Antimicrobial Substances from Fungi, Actinomycetes, Bacteria, and Plants* (Oxford: Oxford University Press, 1949).

Florey, Howard, and Hugh Cairns, *Investigation of War Wounds, Penicillin: A Preliminary Report to the War Office and the Medical Research Council on Investigations Concerning the Use of Penicillin in War Wounds* (London: War Office, 1943).

Foltz, E. L., and H. Wallick, 'Pharmacodynamics of phosphonomycin after intravenous administration in man', *Antimicrobial Agents and Chemotherapy* 9 (1969): 316–21.

Foltz, E. L., H. Wallick, and C. Rosenblum, 'Pharmacodynamics of phosphonomycin after oral administration in man', *Antimicrobial Agents and Chemotherapy* 9 (1969): 322–6.

Fresquet Febrer, José Luis, 'El 606 en España. Una historia de seis meses contada por la prensa', *Anales de la real Academia de Medicina de la Comunitat Valenciana* 12 (2011): 1–23.

Fromtling, R. A., 'Fosfomycin (MK0955): An overview', *Drugs of Today* 23 (1987): 151–8.

Gadebusch, Hans H., Edward O. Stapley, and Sheldon B. Zimmerman, 'The discovery of cell wall active antibacterial antibiotics', *Critical Reviews in Biotechnology* 12 (1992): 225–43.

Galambos, Louis, and Jane Eliot Sewell, *Networks of Innovation: Vaccine Development at Merck, Sharp and Dohme, and Mulford, 1895–1995* (Cambridge: Cambridge University Press, 1997).

Gallego Méndez, Teresa, *Mujer. Falange y franquismo* (Madrid: Taurus, 1983).

Gallego, Antonio, A. Rodríguez, and J. Martínez Mata, 'Fosfomycin: Pharmacological studies', *Drugs Today* 10 Supplement (1974): 161–8.

Gallego, Antonio, and J. M. Rubio, eds., 'Fosfomycin. Proceedings of the International Symposium on Fosfomycin, Madrid, 10 11 July 1975', *Chemotherapy* 23 (1977): Supplement 1.

Gálvez, Lina, *La Compañía Arrendataria de Tabacos, 1887–1945: cambio tecnológico y empleo femenino* (Madrid: LID Editorial Empresarial, 2001).

García Delgado, J. L., 'La industrialización y el desarrollo económico de España durante el franquismo', in *La economía española en el siglo XX. Una perspectiva histórica*, eds. Jordi Nadal, Albert Carreras and Carles Sudriá (Barcelona: Ariel, 1987), 164–89.

Gardner, A. D., 'Jean Orr-Ewing. Born 28th April 1897. Died 17th November 1944', *The Journal of Pathology* 58 (1946): 149.

Gardner, A. D., 'Morphological effects of penicillin on bacteria', *Nature* 146 (1940): 837.

Garrido, Luis, *Las dos biografías de la mujer en España* (Madrid: Instituto de la Mujer, 1993).

Garrod, L. P., 'Mary Barber. 3 April 1911–11 September 1965', *The Journal of Pathology and Bacteriology* 92(2) (1966): 603–10.

Gaudillière, Jean-Paul, 'Circulating mice and viruses: The Jackson Memorial Laboratory, the Cancer Institute and the genetics of breast cancer, 1930–1965', in *The Practices of Human Genetics*, eds. Michael Fortun and Everett Mendelsohn (Dordrecht: Kluwer, 1999).

Gaudillière, Jean-Paul, 'Professional or industrial order? Patents, biological drugs, and pharmaceutical capitalism in early twentieth century Germany', *History and Technology* 24 (2008): 107–33.

Gaudillière, Jean-Paul, 'The singular fate of industrial screening in twentieth century pharmacy: Some thoughts about drugs standardisation and drug regulation', in *Harmonizing Drugs: Standards in 20th Century Pharmaceuticals*, eds. C. Bonah, C. Masutti, Anne Rasmussen and J. Simon (Paris: Glyphe, 2009), 153–80.

Gaudillière, Jean-Paul, and Bernd Gausmeier, 'Moulding national research systems: The introduction of penicillin to Germany and France', *Osiris* 20 (2005): 180–202.

Gaudillière, Jean-Paul, and Ulrike Thoms, eds., *Pharmaceutical Firms and the Construction of Drug Markets: From Branding to Scientific Marketing*, Special Issue of *History and Technology* 29(2) (2013).

Gaudilliére, Jean-Paul, and Volker Hess, eds., *Ways of Regulating Drugs in the 19th and 20th Centuries* (London: Palgrave Macmillan, 2012).

Gaudillière, Jean-Paul, *Inventer la biomédecine: la France, l'Amérique et la production des savoirs du vivant, 1945–1965* (Paris: La découverte, 2002).

Geison, Gerald L., 'The protoplasmic theory of life and the vitalist–mechanist debate', *Isis* 60 (1969): 273–92.

Gewurtz, M. S., 'Looking for Jean Daw: Narratives of women and missionary medicine in modern China', in *Figuring It Out. Science, Gender and Visual Culture*, eds. A. B. Shteir and B. Lightman (Hanover, NH: Darmouth College Press, 2006).

Ghuysen, Jean-Marie, 'The concept of the penicillin target from 1965 until today: The thirteenth Marjory Stephenson Memorial Lecture', *Journal of General Microbiology* 101 (1977): 13–33.

Godin, Benoit, 'Outline for a history of scientific measurement', *Science, Technology and Human Values* 29 (2002): 3–29.

González Blasco, Pedro, *El investigador científico en España* (Madrid: Centro de Investigaciones Sociológicas, 1980).

González, Juan P., Jaime Suñer Pi, and Francisco González, 'Obtención de la penicilina y otras experiencias', *Medicina Clínica* 2(6) (1944): 473–82.

Gradmann, Christoph, 'Biomedical sciences', *Studies in History and Philosophy of Biological and Biomedical Sciences* 30 (2017): 1–5.

Gradmann, Christoph, 'Magic bullets and moving targets: Antibiotic resistance and experimental chemotherapy, 1900–1940', *Dynamis* 31 (2011): 29–46.

Gradmann, Christoph, 'Re-inventing infectious disease: Antibiotic resistance and drug development at the Bayer company, 1945–80', *Medical History* 60 (2016): 155–80.

Gradmann, Christoph, 'Sensitive matters: The World Health Organisation and antibiotic resistance testing, 1945–1975', *Social History of Medicine* 26 (2013): 555–74.

Gradmann, Christoph, and Jonathan Simon, eds., *Evaluating and Standardizing Therapeutic Agents, 1890–1950* (Basingstoke: Palgrave Macmillan, 2010).

Grandin, Karl, Nina Wormbs, and Sven Widmalm, eds., *The Science–Industry Nexus. History, Policy, Implications* (Sagamore Beach, MA: Science History Publication, 2004).

Greene, Jeremy, and Elizabeth S. Watkins, eds., *Prescribed: Writing, Filling, Using and Abusing the Prescription in Modern America* (Baltimore: Johns Hopkins University Press, 2012).

Grote, Mathias, and Maureen A. O'Malley, 'Enlightening the life sciences: The history of halobacterial and microbial rhodopsin research', *FEMS Microbiology Reviews* 35 (2011): 1082–99.

Grote, Mathias, 'Purple matter, membranes and "molecular pumps" in rhodopsin research (1960s–1980s)', *Journal of the History of Biology* 46 (2013): 331–68.

Grote, Mathias, 'Surfaces of action: Cells and membranes in electrochemistry and the life sciences', *Studies in History and Philosophy of Biological and Biomedical Sciences* 41 (2010): 183–93.

Guillamet, Joan, *Tots hem fet estraperlo* (Barcelona: Dima, 1968).

Guirao, Fernando, *Spain and the Reconstruction of Western Europe, 1945–57. Challenge and Response* (London and New York: Macmillan, St. Martin's Press, 1998).

Harris, Dayle A., and Myrle L. Ruger, 'Microbiological aspects of new antibiotic screening. A plug test procedure', *Antibiotics and Chemotherapy* 3 (1953): 265–70.

Health for all beyond 2000. Declaration of Alma-Ata. International Conference on Primary Health Care, Alma-Ata, USSR, 6–12 September 1978, at www.who.int/hpr/NPH/docs/declaration_almaata.pdf.

Hecht, Gabrielle, ed., *Entangled Geographies: Empire and Technopolitics in the Global Cold War* (Cambridge, MA: MIT Press, 2011).

Hendlin, D., E. O. Stapley, M. Jackson, H. Wallik, A. K. Miller, F. J. Wolf, T. W. Miller, L. Chaiet, F. M. Kahan, E. L. Foltz, H. B. Woodruff, J. M. Mata, S. Hernández and S. Mochales, 'Phosphonomycin: A new antibiotic produced by strains of streptomyces', *Science* 166 (1969): 122–3.

Hendlin, D., E. O. Stapley, Sagrario Mochales, and J. Martínez Mata, assignors to Merck & Co., Inc., Rahway, NJ. Antibacterial composition containing (-) (cis-1,2-epoxipropyl) phosphoric acid, United States Patent Office. Patented February 1, 1972.

Hernández, S., Sagrario Mochales, J. Martínez Mata and A. Gallego, 'Un programa de investigación de nuevos antibióticos. I. Resultados del año 1957', *Anales del Instituto de Farmacología Española* VII (1958): 263–6.

Hobby, Gladys L., 'Antimicrobial agents and chemotherapy', *ASM News* 40 (1974): 327–9.

Hobby, Gladys, *Penicillin: Meeting the Challenge* (New Haven: Yale University Press, 1985).

Holloway, W., J. Clark, and R. Rocco, 'Preliminary clinical trials with phosphomycin', *Antimicrobial Agents and Chemotherapy* 9 (1969): 327–31.

Hudson, Pat, Maxine Berg, and Michael Sonenscher, eds., *Manufacture in Town and Country Before the Factory* (Cambridge: Cambridge University Press, 1983).

Hulverscheidt, Marion A., 'The scientific entrepreneur or financing in pharmaceutical research. A portrait of the mariologist Werner Schulemann, 1888–1975', in *Perspectives on Twentieth Century Pharmaceuticals*, eds. Vivian Quirke and Judy Slinn (Oxford: Peter Lang, 2010), 121–48.

Hüntelmann, Axel C., 'Priority, property and trust: Patent laws and pharmaceuticals in the German Empire', *Interdisciplines* 2 (2012): 194–226.

Hurtley, Jacqueline, *Walter Starkie. An Odyssey* (Dublin: Four Courts Press, 2013).

Ignaciuk, Agata 'Publicidad farmacéutica y género: Anuncios de la píldora en España (1964–1985)', *Medicina e Historia* 1 (2014): 4–21.

Jacoby, G. A., 'History of antimicrobial agents and chemotherapy from 1972 to 1998', *Antimicrobial Agents and Chemotherapy* 43 (1999): 999–1002.

Jiménez Casado, Mariano, *Doctor Jiménez Díaz. Vida y obra* (Madrid: Fundación Conchita Rábago de Jiménez Diaz, 1993).

Jiménez Díaz, Carlos 'Significación conceptual del conocimiento de la endocarditis bacteriana', *Revista Clínica Española* 70 (1958): 277–9.

Jiménez Díaz, Carlos, 'Observaciones sobre el empleo de los antibióticos', *Revista Clínica Española* 47(3) (1952): 143–53.

Jimenez Lucena, Isabel, 'El tifus exantemático de la posguerra española (1939–1943). El uso de una enfermedad colectiva en la legitimización del Nuevo Estado', *Dynamis* 14 (1994): 185–98.

Jimenéz Quesada, Mateo, *De Fleming a Marañón. Anécdotas de mi archivo* (Madrid: Minaza, 1972).

Jiménez Quesada, Mateo, *El tratamiento de las infecciones de garganta, nariz, oídos y vías bajas respiratorias con nebulizadores de antibióticos* (Madrid: Faure, 1947).

Jordanova, Ludmila J., 'Natural facts: A historical perspective on science and sexuality', in *Nature, Culture and Gender*, eds. C. P. MacCormack and M. Strathern (Cambridge: Cambridge University Press, 1992), 42–69.

Jordanova, Ludmilla J., 'Gender and the historiography of science', *British Journal for the History of Science* 26 (1993): 469–83.

Kahan, Frederick M., Jean S. Kahan, Patrick J. Cassidy, and Helmut Kropp, 'The mechanism of action of fosfomycin (phosphonomycin)', *Annals of the New York Academy of Sciences* 235 (1974): 364–87.

Kaplan, T., 'Social movements of women and the public good', in *Dones en movement(s)*, eds. C. Borderías and M. Renom (Barcelona: Icaria, 2008), 19–47.

Kapur, Akash, 'Rush', *The New Yorker*, October 14, 2013.

Keating, Peter, and Albert Cambrosio, *Biomedical Platforms: Realigning the Normal and the Pathological in Late-Twentieth-Century Medicine* (Cambridge, MA: MIT Press, 2003).

Keefer, Chester S., 'The present status of penicillin in the treatment of infections', *Proceedings of the American Philosophical Society* 88 (1944): 174–6.

Kingston, William, 'Streptomycin, Schatz v. Waksman, and the balance of credit for discovery', *Journal of the History of Medicine and Allied Sciences* 59 (2004): 441–62.

Kohler, Robert E., 'The enzyme theory and the origin of biochemistry', *Isis* 64 (1973): 181–96.

Kohler, Robert E., *Lords of the Fly: Drosophila Genetics and the Experimental Life* (Chicago: University of Chicago Press, 1994).

Kojevnikov, Alexei, 'The making of the Soviet bomb and the shaping of Cold War science', in *Reappraising Oppenheimer: Centennial Studies and Reflections*, eds. C. Carson and D. A. Hollinger (Berkeley: Office for History of Science and Technology, 2005), 129–45.

Krige, John, *American Hegemony and the Postwar Reconstruction of Science in Europe* (Cambridge, MA: MIT Press, 2006).

Kutz Echave, R., 'El empleo de la penicilina en la práctica médica', *Revista Clínica Española* 19(5) (1945): 340–9.

Landecker, Hanna, *Culturing Life: How Cells Became Technologies* (Cambridge, MA: Harvard University Press, 2007).

Lechevalier, Hubert, 'The search for antibiotics at Rutgers University', in *The History of Antibiotics. A Symposium*, ed. John Parascandola (Madison: American Institute of the History of Pharmacy, 1980), 113–23.

Lederberg, Joshua, 'Bacterial protoplasts induced by penicillin', *Proceedings of the National Academy of Sciences of the United States of America* 42 (1956): 574–8.

Lederberg, Joshua, 'Mechanism of action of penicillin', *Journal of Bacteriology* 73 (1957): 144.

Lederberg, Joshua, 1958. Nobel lecture: A view of genetics. *Nobelprize.org*. Nobel Media, at www.nobelprize.org/nobel_prizes/medicine/laureates/1958/lederberg-lecture.html.

Lederberg, Joshua, and Jacqueline St. Clair, 'Protoplasts and L-type growth of *Escherichia coli*', *Journal of Bacteriology* 75 (1958): 143–60.

Lesch, John E., *The First Miracle Drugs: How the Sulfa Drugs Transformed Medicine* (New York: Oxford University Press, 2007).

Ley Gracia, Adolfo, *Pasado, presente y future de la cirugía craneocerebral.* Discurso leido por el académico electo en el acto de su recepción (Barcelona: Real Academia de Medicina de Barcelona, 1971).

Lie, Anne Kveim, 'Producing standards, producing the Nordic region: Antibiotic susceptibility testing, from 1950–1970', *Science in Context* 27 (2014): 215–48.

Liebenau, Jonathan, 'The British success with penicillin', *Social Studies of Science* 17 (1987): 69–86.

Lindee, M. Susan, 'Keeping secrets, learning to lie', paper presented at Dark Matters: Contents and Discontents of Cold War Science International Conference, Barcelona, May 31–June 2, 2013.

Lobo, Félix, *Medicamentos: Política y Economía* (Barcelona: Masson S.G. 1992).

López García, Santiago, 'El Patronato Juan de la Cierva (1939–1960). Parte II. La investigación científica y tecnológica', *Arbor* 162 (2013): 1–32.

López, Santiago, and Mar Cebrián, 'Economic growth, technology transfer, and convergence in Spain, 1960–1973', in *Technology and Human Capital in Historical Perspective*, eds. Jonas Ljunberg and Jan-Pieter Smits (New York: Palgrave Macmillan, 2004), 120–44.

López, R., 'Comportamiento de algunas especies de azotobacteriáceas frente a los antibióticos', *Microbiología Española* 20 (1967): 167–90.

López, R., E. Ronda, and A. Portolés, 'Variaciones en la composición de aminoácidos en especies de azotobacterias antibióticorresistentes', *Microbiología Española* 20 (1967): 203–18.

Łotysz, Sławomir, 'Controlling the production and distribution of drugs in communist Poland', *Medicina nei Secoli: Arte e Scienza* 26 (2014): 519–56.

Löwy Ilana, and Jean-Paul Gaudillière, eds., *Invisible Industrialist. Manufactures and the Production of Scientific Knowledge* (London: Macmillan, 1998).

Löwy, Ilana, and H. Rouch, 'Genèse et development du genre: les sciences et les origins de la distinction entre sexe et genre', *Cahiers du Genre* 34 (2003): 5–16.

Löwy, Ilana, *Between Bench and Bedside: Science, Healing, and Interleukin-2 in a Cancer Ward* (Cambridge, MA: Harvard University Press, 1996).

Löwy, Ilana, *Preventive Strikes. Women, Precancer and Prophylactic Surgery* (Baltimore: Johns Hopkins University Press, 2010).

Macfarlane, Gwyn, *Alexander Fleming. The Man and the Myth* (Cambridge, MA: Harvard University Press, 1984).

Mahoney, Tom, *The Merchants of Life: An Account of the American Pharmaceutical Industry* (New York: Harper, 1959).

Marañón, Gregorio, 'Comentarios sobre un caso de meningitis estafilocócica', *Medicamenta* 3(80) (June 1945): 361–2.

Marks, Harry M., *The Progress of Experiment: Science and Therapeutic Reform in the United States, 1900–1990* (Cambridge: Cambridge University Press, 2000).

Marks, Lara, *Sexual Chemistry: A History of the Contraceptive Pill* (New Haven: Yale University Press, 2010).

Martín-Aceña, Pablo, and Francisco Comín, *INI. 50 años de industrialización en España* (Madrid: Espasa Calpe, 1991).

Martin, H. H., 'Bacterial protoplasts—A review', *Journal of Theoretical Biology* 5 (1963): 1–34.

Martínez Mata, J., 'La investigación de antibióticos en la industria farmacéutica española. Historia de un descubrimiento: la fosfomicina', *Química e Industria* 20 (1974): 51–9.

Maurois, André, *La vie de Sir Alexander Fleming* (Paris: Hachette, 1959).

McQuillen, Kenneth, and Milton R. J. Salton, 'Synthetic media for maintenance and induction of lysogenic *Bacillus megaterium*', *Biochimica et Biophysica Acta* 16 (1955): 596–7.

McQuillen, Kenneth, 'Bacterial protoplasts: Growth and division of protoplasts of *Bacillus megaterium*', *Biochimica et Biophysica Acta* 18 (1955): 458–61.

McQuillen, Kenneth, 'Bacterial protoplasts', in *The Bacteria: A Treatise on Structure and Function*, eds. I. C. Gunsalus and R. Y. Stanier (New York and London: Academic Press, 1960), 249–359.

Medina García, E., 'Contrabando en la frontera de Portugal: orígenes, estructuras, conflicto y cambio social' (PhD diss., Universidad Complutense de Madrid, Madrid, 2004).

Medina-Doménech, Rosa, and Alfredo Menéndez-Navarro, 'Cinematic representations of medical technologies in the Spanish official newsreel, 1943–1975', *Public Understanding of Science* 14 (2005): 393–408.

Miller, K., 'Antibiotic MSD-235. III evaluation of antibiotic MSD-235 in vitro and in experimental infections in mice', *Antimicrobial Agents and Chemotherapy* (1969): 33–7.

Mir, Conxita, Carmen Agustí, and Josep Gelonch, eds., *Pobreza, marginación, delincuencia y políticas sociales bajo el franquismo* (Lleida: Universitat de Lleida, 2014).

Mokyr, Joel, 'The rise and fall of the factory system: Technology, firms and households since the industrial revolution', *Carnegie-Rochester Conference Series on Public Policy* 55 (2001): 1–45.

Moradiellos, Enrique, *La España de Franco, 1939–1975: política y sociedad* (Madrid: Síntesis, 2000).

Muñoz Ruiz, Emilio, and Julio Rodriguez-Villanueva, 'Un método rápido y sencillo para el aislamiento de nuevos microorganismos productores de antibióticos', *Microbiología Española* 13 (1960): 409–14.

Muñoz, E., M. Victoria Elorza, and J. Rodríguez-Villanueva, 'Clasificación de un microorganismo de acción lítica como variedad de *Streptomyces violaceus*', *Microbiología Española* 17(4) (1964): 195–208.

Muñoz, Emilio, and Florencio Ornia, *Ciencia y tecnología: una oportunidad par España* (Madrid: Ministerio de Educación y Ciencia-Aguilar, 1986).

Navarro, Vicente, 'A critique of the ideological and political positions of the Willy Brandt report and the WHO Alma Ata declaration', *Social Science and Medicine* 18 (1984): 467–74.

Ndiaye, Pap, *Nylon and Bombs: DuPont and the March of Modern America* (Baltimore: Johns Hopkins University Press, 2007).

Neushul, Peter, 'Science, government and the mass production of penicillin', *Journal of the History of Medicine and Allied Sciences* 48 (1993): 371–95.

Olagüe de Ros, Guillermo, *Del uso de la retórica en el discurso científico: a propósito de los programas de trabajo de Fidel Fernández Martínez (1890–1942) y Eduardo Ortiz de Landázuri (1910–1985). Discurso leído en el acto de recepción pública* (Granada: Real Academia de Medicina y Cirugía de Granada, 2001).

Ortega Núñez, A., 'Terramicina (un nuevo antibiótico)', *Revista Clínica Española* 40 (1951): 332–9.

Ortiz de Landázuri, Eduardo, and J. Perianes, 'Normas generales de la terapéutica penicilínica', *Revista Clínica Española* 17(5) (1945): 361–72.

Ortiz Gómez, Teresa, *Medicina, historia y género. 130 años de investigación eminista* (Oviedo: KRK, 2006).

Ortiz Gómez, Teresa, and M. J. Santesmases, eds., *Gendered Drugs and Medicine Historical and Socio-Cultural Perspective* (Farham: Ashgate Academic Publishers).

Oudshoorn, Nelly, *Beyond the Natural Body: An Archaeology of Sex Hormones* (London: Routledge, 2003).

Parascandola, John, 'John Mahoney and the introduction of penicillin to treat syphilis', *Pharmacy in History* 43 (2001): 3–13.

Parascandola, John, 'The introduction of antibiotics into therapeutics', in *Studies in the History of Modern Pharmaceutical and Drug Therapy* (Farham: Ashgate 2012), 261–80.

Parascandola, John, ed., *The History of Antibiotics: A Simposium* (Madison: American Institute of the History of Pharmacy, 1980).

Park, James T., and Jack L. Strominger, 'Mode of action of penicillin', *Science* 125 (1957): 99–101.

Pau-Roca, L., and J. Coll-Sampol, 'Terapéutica antibióticas, bacteriológicamente dirigida, en asmología', *Revista Clínica Española* 46 (1957): 313–22.

Paul, D. B., and J. P. Brosco, *The PKU Paradox: A Short History of a Genetic Disease* (Baltimore: Johns Hopkins University Press, 2013).

Peláez, Fernando, 'Introduction: 50 years of CIBE', in *CIBE: 50 años de investigación* (Madrid: Merck Sharp and Dhome España, 2005), 23–33, on 24–5.

Peláez, Fernando, 'The historical delivery of antibiotics from microbial natural products—Can history repeat?' *Biochemical Pharmacology* 71 (2006): 981–90.

Peña González, Beatriz, *El estraperlo en Cádiz: la realidad social* (Cádiz: Quorum, 2004), 132–86.

Pérez Gorricho, María Beatriz, 'Farmacovigilacia de los antibióticos. Estructura de consumo y mecanismos de control' (PhD diss., Madrid: Universidad Complutense, 1985).

Pérez-Ureña, María Teresa, 'Antibioticorresistencia in vitro de estafilococos tratados por asociaciones de penicilina con neomicina, novobiocina y kanamicina', *Microbiología Eespañola* 19 (1966): 287–98.

Pérez-Ureña, María Teresa, 'Valoración de la antibioticorresistencia de bacilos gram-negativos sometidos a interacciones antibacterianas de carácter sinérgico', *Microbiología Española* 19 (1966): 299–320.

Podolsky, Scott H., 'Antibiotics and the social history of the controlled clinical trial', *Journal for the History of Medicine and Allied Sciences* 65 (2010): 327–67.

Podolsky, Scott H., *The Antibiotic Era: Reform, Resistance, and the Pursuit of a Rational Therapeutics* (Baltimore: Johns Hopkins University Press, 2014).

Podolsky, Scott H., Robert Bud, Christoph Gradmann, Bård Hobaek, Claas Kirchhelle, Tore Mitvedt, María Jesús Santesmases, Ulrike Thoms, Dag Berild and Anne Kveim Lie, 'History teaches us that confronting antibiotic resistance requires stronger global collective action', *Journal of Law, Medicine & Ethics*, Special Supplement to vol. 4 (2015): 27–32.

Porras Gallo, Isabel, 'Antecedentes y creación del Instituto de Sueroterapia, Vacunación y Barteriología de Alfonso XIII', *Dynamis* 18 (1998): 81–105.

Porras Gallo, Isabel, 'La lucha contra las enfermedades "evitables" en España y la pandemia de la gripe de 1918–19', *Dynamis* 14 (1994): 159–83.

Porras Gallo, Isabel, *Un reto para la sociedad madrileña: la epidemia de gripe de 1918–19* (Madrid: Editorial Complutense, 1997).

Porras Gallo, Isabel, Maria José Báguena and Mariano Ayarzagüena, eds., *La erradicación y el control de las enfermedades infecciosas* (Madrid: Catarata, 2016).

Porras Gallo, Isabel, Mariano Ayarzagüena Sanz, Jaime de las Heras Salord, María José Báguena Cervellera, eds., Rosa Ballester Añón, Inês Guerra Santos, Juan Antonio Rodríguez Sánchez, and José Vicente Toledo Marhuenda, *El drama de la polio. Un problema social y familiar en la España franquista* (Madrid: Los libros de la Catarata, 2013).

Portolés, Antonio, and Enrique Feduchi, 'Micosis urinarias subordinadas a la presencia de sustancias antibióticas', *Microbiología Española* 12(3) (1959): 199–242.

Portolés, Antonio, '¿Intervienen los antibióticos en la formación de levaduras?' *Medicamenta* 10 (1953): 53–5.

Puerto Sarmiento, F. Javier, *Historia de la Real Academia Nacional de Farmacia* (Madrid: Real Academia Nacional de Farmacia, 2012).

Puig, Nuria, 'El crecimiento asistido de la industria química en España: Fabricación Nacional de Colorantes y Explosivos, 1922–1965', *Revista de Historia Industrial* 15 (1999): 105–36.

Puig, Nuria, and Javier Loscertales, 'Las estrategias de crecimiento de la industria química alemana en España, 1880–1936: exportación e inversión directa', *Revista de Historia Económica* 19 (2001): 345–82.

Quirke, Viviane, 'Penicillin and post-war reconstruction in Britain and France: Science and industry in the balance', *La letter de la Maison Française d'Oxford* 6 (1997): 87–98.

Quirke, Vivian, and Judy Slinn, eds., *Perspectives on Twentieth Century Pharmaceuticals* (Oxford: Peter Lang, 2010).

Radcliff, John Druty, *Yellow Magic. The Story of Penicillin* (New York: Random House, 1945).

Rentetzi, Maria, 'Packaging radium, selling science: Boxes, bottles and other mundane things in the world of science', *Annals of Science* 68 (2000): 375–99.

Rentetzi, Maria, *Trafficking Materials and Gendered Experimental Practices: Radium Research in Early 20th Century Vienna* (New York: Columbia University Press, 2008).

Rentetzi, Maria, 'The tobacco museum of the city of Kavala', *Technology and Culture* 50(3) (2009): 649–57.

Rheinberger, Hans-Jörg, 'Comparing experimental systems: Protein synthesis in microbes and in animal tissue at Cambridge (Ernest F. Gale) and at the Massachusetts General Hospital (Paul C. Zamecnik), 1945–1960', *Journal of the History of Biology* 29 (1996): 387–416.

Rheinberger, Hans-Jörg, *Toward a History of Epistemic Things: Synthesizing Proteins in the Test Tube* (Stanford: Stanford University Press).

Richards, Alfred N., 'Penicillin: A statement released by the committee of medical research', *JAMA* 122 (1943): 235.

Rodríguez Barreira, Óscar, 'Cambalaches: hambre, moralidad popular y mercados negros de guerra y postguerra', *Historia Social* 77 (2013): 149–74.

Rodríguez-Ocaña, Esteban, 'The politics of public health in state-managed schemes of healthcare in Spain (1940–1990)', in *Images of Disease: Science, Public Policy and Health in Post-war Europe*, eds. Ilana Löwy and John Krige (Luxembourg: Office for Official Publications of the European Communities, 2001), 187–210.

Rodríguez Ocaña, Esteban, and Ferrán Martínez Navarro, *Salud pública en España: de la Edad Media al siglo XXI* (Granada: Escuela Andaluza de Salud Pública, 2008).

Romero de Pablos, Ana, 'Patents, antibiotics, and autarky in Spain', *Medicina nei secoli* 26(2) (2013): 423–49.

Romero de Pablos, Ana, 'Regulation and the circulation of knowledge: Penicillin patents in Spain', *Dynamis* 31(2) (2011): 363–83.

Rossiter, Margaret, *Women Scientists in America: Before Affirmative Action 1940–1972* (Baltimore: Johns Hopkins University Press, 1995).

Sakula, Alex, 'Selman Waksman 1888–1973, discoverer of streptomycin: A centenary review', *British Journal of Diseases of the Chest* 82 (1988): 23–31.

Salton, Milton R. J., 'Cell structure and the enzymic lysis of bacteria', *Journal of General Microbiology* 9 (1953): 512–23.

Salton, Milton R. J., 'Cell wall of *Micrococcus lysodeikticus* as the substrate of lysozyme', *Nature* 170 (1952): 746.

Salton, Milton R. J., and Kenneth McQuillen, 'Bacterial protoplasts: II. Bacteriophage multiplication in protoplasts of sensitive and lysogenic strains strains of *Bacillus megaterium*', *Biochimica et Biophysica Acta* 17 (1955): 465–72.

San Román, Elena, *Ejército e industria: el nacimiento del INI* (Barcelona: Crítica, 1999).

Sánchez Agustí, F., *Espías, contrabando, maquis y evasión: la Segunda Guerra Mundial en los Pirineos* (Lleida: Milenio, 2010).

Sánchez Ron, José Manuel, 'Política científica e ideología: Albareda y los primeros años del Consejo Superior de Investigaciones Científicas', *Boletín de la Institución Libre de Enseñanza* 14 (1992): 53–74.

Sánchez, J. G., and E. G. Sánchez, 'Antibióticos y cine: El tercer hombre y Mercado prohibido', *Revista Española de Quimioterapia* 17 (2004): 223–5.

Santesmases, M. J., *Entre Cajal y Ochoa: ciencias biomédicas en la España de Franco* (Madrid: CSIC, 2001).

Santesmases, M. J., 'From prophylaxis to atomic cocktail: Circulation of radioiodine', *Dynamis* 29 (2009): 337–63.

Santesmases, M. J., 'Orígenes internacionales de la política científica', in *Cien años de política científica en España*, eds. Ana Romero and M. J. Santesmases (Madrid: Fundación Banco Exterior, 2008), 293–327.

Santesmases, M. J., *Mujeres científicas en España: profesionalización y modernización social* (Madrid: Instituto de la Mujer, 2000).

Santesmases, M. J., 'Productividad, diplomacia y dependencia tecnica: España en Occidente durante el franquismo', *Pensamiento Iberoamericano* 5 (2009): 21–41.

Santesmases, M. J., 'Severo Ochoa and the biomedical sciences in Spain under Franco, 1959–1975', *Isis* 91 (2000): 706–34.

Santesmases, María Jesús, and Christoph Gradmann, 'Circulation of antibiotics: An introduction', *Dynamis* 31(2) (2011): 293–303.

Santesmases, M. J., and E. Suárez-Díaz, 'A cell-based epistemology: Human genetics in the era of biomedicine', *Historical Studies in the Natural Sciences* 45 (2015): 1–13.

Santesmases, M. J., 'Distributing penicillin: The clinic, the hero and industrial production in Spain, 1943–1952', in *Perspectives on Twentieth-Century Pharmaceuticals*, ed. Viviane Quirke and Judy Slinn (Oxford-Berna: Peter Lang, 2010), 91–118.

Santesmases, M. J., 'Screening antibiotics: Industrial research by CEPA and Merck in the 1950s', *Dynamis* 31 (2011): 407–427.

Santesmases, M. J., 'Gender in research and industry: Women in antibiotic factories in 1950s Spain', in *Gendered Drugs and Medicine: Historical and Socio-cultural Perspectives*, ed. T. Ortiz-Gómez and M. J. Santesmases (Farnham: Ashgate, 2014), 61–84.

Santesmases, M. J., 'The long postwar and the politics of penicillin: Early circulation and smuggling in Spain, 1944–1954', *Medicina nei Secoli* 26 (2014): 615–38.

Santesmases, M. J., Antibióticos en la Autarquía (Madrid: Fundación Empresa Pública, 1999).

Santesmases, M. J., 'The bacterial cell wall in the antibiotic era: An ontology in transit between morphology and metabolism, 1940s–1960s', *Journal of the History of Biology* 49 (2016): 3–36.

Sarasúa, Carmen, and Lina Gálvez, eds., *¿Privilegios o eficiencia? Mujeres y hombres en los mercados de trabajo* (Alicante: Publicaciones de la Universidad de Alicante, 2003).

Saz, Ismael, 'El primer franquismo', *Ayer* 36 (1999): 201–21.

Saz, Ismael, *España contra España: Los nacionalismos franquistas* (Madrid: Marcial Pons, 2003).

Saz, Ismael, *Las caras del franquismo* (Granada: Comares, 2013).

Schaffer, Simon, Lissa Roberts, Kapil Raj, and James Delbourgo, eds., *The Brokered World: Go-betweens and Global Intelligence, 1770–1820* (Sagamore Beach, MA: Science History Publications, 2009).

Schatz, Albert, 'Some personal reflections on the discovery of streptomycin', *Pakistan Dental Review* 15 (1965): 125–34.

Scott, Joan, 'Gender, a useful category in historical analysis', *American Historical Review* 91 (1986): 1053–75.

Secord, James A., *Victorian Sensation: The Extraordinary Publication, Reception, and Secret Authorship of Vestiges of the Natural History of Creation* (Chicago: University of Chicago Press, 2003).

Segovia de Arana, José M., and Juan M. de Palacios Mateos, *Estreptomicina* (Madrid: Paz Montalvo, 1947).

Segura, Cristina, 'Mujeres, trabajo y familia en las sociedades industriales', in *La historia de las mujeres: una revisión historiográfica*, eds. Isabel del Val Valdivieso, Magdalena S. Tomás Pérez, Jesús Dueñas Cepeda, Cristina de la Rosa Cubo (Valladolid: Universidad de Valladolid-AEIHM, 2004).

Serralonga i Urquidi, Joan, 'El cuento de la regularización sanitaria y asistencial en el Régimen franquista. Una primera etapa convulsa, 1936–1944', *Historia Social* 59 (2007): 77–98.

'Sesión necrológica en homenaje al Excmo. Señor Don Antonio Portolés Alonso', *Anales de la Real Academia nacional de Farmacia* 72 (2006): 165–92.

Sevillano Calero, F., *Propaganda y medios de comunicación en el franquismo* (Alicante: Universidad de Alicante, 1998), at http://publicaciones.ua.es/filespubli/pdf/LD84790874637062078.pdf.

Silver, Lynn L., 'Rational approaches to antibacterial discovery: Pre-genomic directed and phenotypic screening', in *Antibiotic Discovery and Development*, eds. T. J. Dougherty and M. J. Pucci (New York: Springer, 2012), 33–75.

Smith, J. W., and J. P. Sanford, 'Activity of phosphonomycin in Nasal carriers of coagulase positive staphylococci', *Antimicrobial Agents and Chemotherapy* 9 (1969): 346–8.

Smith, L. D., and Telma Hay, 'The effect of penicillin on the growth and morphology of *Staphylococcus aureus*', *The Journal of the Franklin Institude* 233 (1942): 598–602.

Snelders, Stephen, Charles Kaplan, and Toine Pieters, 'On cannabis, chloral hydrate, and career cycles of psychotrophic drugs in medicine', *Bulletin of the History of Medicine* 80 (2006): 95–114.

Somovilla, Miguel, *Retrato de David Vázquez 1930–1986. De la sidra a los antibióticos* (La Felguera: Comisión Homenaje a David Vázquez).

Spellberg, B., 'Dr. W. H. Stewart, mistaken or maligned?' *Clinical Infectious Disease* 47 (2008): 294.

Stapley, E. O., D. Hendlin, J. M. Mata, M. Jackson, H. Wallick, S. Hernández, S. Mochales, S. A. Currie, and R. M. Miller, 'Phosphonomycin I. Discovery and in vitro biological characterization', *Antimicrobial Agents and Chemotherapy* 9 (1968): 284–90.

Stapley, E. O., J. M. Mata, I. M. Miller, T. C. Demny, and H. B. Woodruff, 'Antibiotic MSD-235. I. Production by *Streptomices avidinii* and *Streptomyces lavandulae*', *Antimicrobial Agents and Chemotherapy* 3 (1963): 20–7.

Stapley, Edward O., 'Cross-resistance studies and antibiotic identification', *Applied Microbiology* 6 (1958): 392–8.

Strasser, Bruno, 'Collecting and experimenting: The moral economies of biological research, 1960–1980s', in *History and Epistemology of Molecular Biology and Beyond: Problems and Perspectives* (Berlin: Max Planck Institut für Wissenschaftsgeschichte, 2006), 105–25.

Strasser, Bruno, 'Collecting nature: Practices, styles, and narratives', *Osiris* 27 (2012): 303–40.

Štrbáňová, Sonia, *Holding Hands with Bacteria: The Life and Work of Marjory Stephenson* (Berlin: Springer, 2016).

Strohl, W. R., H. B. Woodruff, R. L. Monaghan, D. Hendlin, S. Mochales, A. L. Demain, and J. Liesch, 'The history of natural products research at Merck & Co.', *SIM News* 51 (2001): 5–19.

Suárez-Díaz, Edna, Gisela Mateos, and Ana Barahona, 'Across borders: Science and technology during the Cold War: An introduction', *Dynamis* 35 (2015): 271–8.

Summers, William C., 'Microbial drug resistance: A historical perspective', in *Bacterial Resistance to Antimicrobials*, eds. Richard G. Wax, Kim Lewis, Abigail A. Salyers and Harry Taber (Boca Raton: CRC Press, 2008), 1–9.

Swann, John P., *Academic Scientists and the Pharmaceutical Industry: Cooperative Research in Twentieth-century America* (Baltimore: Johns Hopkins University Press, 1988).

Tagliacozzo, E, *Secret Trades, Porous Borders: Smuggling and States Along a Southeast Asian Frontier, 1865–1915* (New Haven, CT: Yale University Press, 2005).

Tansey, E. M., and L. A. Reynolds, *Post-Penicillin Antibiotics: From Acceptance to Resistance*, Witness Seminar (London: Wellcome, 2000).

Tejerina, Genoveva, 'Desarrollo de la resistencia a penicilina, estreptomicina y aureomicina en Mycobacterium phlei y *Escherichia coli*', *Microbiología Española* 13 (1960): 393–408.

Tello Valdivieso, Federico, 'Aportaciones al conocimiento de las micosis broncopulmonares', *Revista Clínica Española* 20 (1946): 1–24.

Thoms, Ulrike, 'Between promise and threat. Antibiotics in foods in West Germany 1950–1980', *NTM Zeitschrift für Geschichte der Wissenschaften, Technik und Medizin* 20(3) (2012): 181–214.

Tone, Andrea, and Elizabeth Siegel Watkins, eds., *Medicating Modern America: Prescription Drugs in History* (New York: New York University Press, 2007).

Townson, Nigel, ed., *Spain Transformed: The Franco Dictatorship, 1959–1975* (Basingstoke: Palgrave Macmillan, 2010).

Trías de Bes, Luis, 'Alexander Fleming y su penicilina', *La Vanguardia*, May 26, 1948.

Trueta, Josep, *Fragments d'una vida. Memories* (Barcelona: Edicions, 1978).

Turchetti, Simone, Nestor Herran, and Soraya Boudia, 'Introduction: Have we ever been "transnational"? Towards a history of science across and beyond borders', *The British Journal for the History of Science* 45 (2012): 319–36.

Tusell, Javier, *Franco y los católicos: la política interior española entre 1945 y 1957* (Madrid: Alianza, 1984).

Vicedo, Marga, 'The father of ethology and the foster mother of ducks: Konrad Lorenz as expert on motherhood', *Isis* 100 (2009): 263–91.

Vilanova, Xavier, 'El tratamiento penicilínico de la sífilis', *Revista Clínica Española* 58 (1955): 307–20.

Villanueva Vadillo, Víctor, *Factores determinantes de un proyecto de industria químico farmacéutica nacional, discurso leído en la toma de posesión como académico de número* (Madrid: Real Academia de Farmacia, 1945).

Viñas, Ángel, *Los pactos secretos de Franco con Estados Unidos: bases, ayuda económica, recortes de soberanía* (Barcelona: Grijalbo, 1981).

Wainwright, Milton, 'Streptomycin: Discovery and resultant controversy', *History and Philosophy of the Life Sciences* 13 (1991): 97–124.

Wainwright, Milton, and Harold T. Swan, 'C.G. Paine and the earliest surviving clinical records of penicillin therapy', *Medical History* 30 (1986): 42–56.

Waksman, Selman A., 'Antagonistic relations of microorganisms', *Bacteriology Reviews* 5 (1941): 231–91.

Waksman, Selman, 'Investigación de nuevo agentes quimioterápicos', *Microbiología Española* 3 (1950): 83–97.

Waksman, Selman, 'Streptomycin: Background, isolation, properties, and utilization', *Nobel Lecture*, December 12 (1952): 370–88.

Waksman, S. A., 'What is an antibiotic or an antibiotic substance?' *Mycologia* 39 (1947): 565–9.

Waksman, Selman A., and H. Boyd Woodruff, 'The soil as a source of microorganisms antagonistic to disease-producing bacteria', *Journal of Bacteriology* 40 (1940): 581–600.

Watanabe, Tsutomu, 'Infective heredity of multiple drug resistance in bacteria', *Bacteriological Reviews* 27 (1963): 87.

Watkins, Elizabeth S., *On the Pill: A Social History of Oral Contraceptives, 1950–1970* (Baltimore: Johns Hopkins University Press, 2011).

Weatherall, Mark, *In Search of a Cure. A History of Pharmaceutical Discovery* (Oxford: Oxford University Press, 1990).

Weidel, Wolfhard, and H. Pelzer, 'Bag-shaped macromolecules. A new outlook on bacterial cell walls', *Advances in Enzymology* 26 (1964): 193–232.

Woodruff, H. B., and L. E. MacDaniel, 'The antibiotic approach', in *The Strategy of Chemistry*, eds. S. T. Cohen and R. Rowatt (Cambridge: Cambridge University Press, 1958), 29–48.

Woodruff, H. B., Oral history interview, 18 August 2004, by Shaun Illingworth and Nicholas Trajano Molnar, p. 36, Rutgers Oral History Archives, at http://oralhistory.rutgers.edu/rutgers-history/31-interviewees/1337-woodruff-h-boyd, accessed June 28, 2015.

Woodruff, H. Boyd, and Selman Waksman, 'Historical background', *Annals of the New York Academy of Sciences* 89 (1960): 287–98.

Woodruff, H. Boyd, Sebastián Hernández, and Edward O. Stapley, 'Evolution of an antibiotic screening programme. A tribute to Justo Martínez Mata', *Hindustan Antibiotics Bulletin* 21 (1979): 71–84.

Worboys, Michael, *Spreading Germs: Disease Theories and Medical Practice in Britain, 1865–1900* (Cambridge: Cambridge University Press, 2000).

Worboys, Michael, and Flurin Condrau, eds., *Tuberculosis Then and Now: Interdisciplinary Perspectives on a Post-modern Plague* (Montreal: McGill University Press, 2010).

World Health Organisation, *The Selection of Essential Drugs* (Geneva: WHO, 1977).

Index

M. J. Santesmases, *The Circulation of Penicillin in Spain*, Medicine and Biomedical Sciences in Modern History,
https://doi.org/10.1007/978-3-319-69718-5

GPSR Compliance
The European Union's (EU) General Product Safety Regulation (GPSR) is a set of rules that requires consumer products to be safe and our obligations to ensure this.

If you have any concerns about our products, you can contact us on

ProductSafety@springernature.com

In case Publisher is established outside the EU, the EU authorized representative is:

Springer Nature Customer Service Center GmbH
Europaplatz 3
69115 Heidelberg, Germany

www.ingramcontent.com/pod-product-compliance
Lightning Source LLC
Chambersburg PA
CBHW040823150926
51501CB00010B/196

* 9 7 8 3 3 1 9 6 9 7 1 7 8 *